RETURN TO CLARK LIBRARY

This book is due on the last date stamped below.

MAR 1 6 1994

NOV 1 7 1994

MAR 2

APR

BASIC BUSINESS STATISTICS FOR MANAGERS

BASIC BUSINESS STATISTICS FOR MANAGERS

Alan S. Donnahoe

John Wiley & Sons, Inc.
New York • Chichester • Brisbane • Toronto • Singapore

Publisher: Stephen Kippur
Editor: Katherine Schowalter
Managing Editor: Ruth Greif
Editing, Design, and Production: Publication Services

This publication is designed to provide accurate and authoritative information in regard to the subject matter covered. It is sold with the understanding that the publisher is not engaged in rendering legal, accounting, or other professional service. If legal advice or other expert assistance is required, the services of a competent professional person should be sought. FROM A DECLARATION OF PRINCIPLES JOINTLY ADOPTED BY A COMMITTEE OF THE AMERICAN BAR ASSOCIATION AND A COMMITTEE OF PUBLISHERS.

Copyright © 1988 by Alan S. Donnahoe

All rights reserved. Published simultaneously in Canada.

Reproduction or translation of any part of this work beyond that permitted by section 107 or 108 of the 1976 United States Copyright Act without the permission of the copyright owner is unlawful. Requests for permission or further information should be addressed to the Permission Department, John Wiley & Sons, Inc.

Library of Congress Cataloging-in-Publication Data
Donnahoe, Alan S.
    Basic business statistics for managers.
    1. commercial statistics. 2. Statistics. I. Title.
HF1017.D64 1988 519.5'024658 87-24273
ISBN 0-471-62939-1
ISBN 0-471-62940-5 (pbk.)

Printed in the United States of America
88 89 10 9 8 7 6 5 4 3 2 1

PREFACE

If you would like to improve your understanding of numbers, their interpretation and application, then this book is for you.

This is useful knowledge for anyone, but especially for executives and professionals in the business world, where numbers are the universal language. If you don't really understand this language, you will be under a definite handicap throughout your career.

I can speak directly to this point based on twenty years' experience as the chief executive of a company that grew during this period from minor size to Fortune 500 status. I know at first hand how executives can benefit from a proper grasp of the subject, in quality of judgment and in overall job performance.

The methods used to analyze numbers are the techniques of statistical analysis. I know this sounds formidable and intimidates people. So your first lesson is to put this prejudice aside.

If you read this book you will find that you don't need to be a mathematician to make use of these techniques. They really are quite easy to understand and to apply.

You will find that they are logical, straightforward, and practical: designed in each case to solve real problems in the real world.

You will discover that many of these techniques are quite ingenious in producing simple answers to difficult questions and in arriving at broad inferences based on a minimum of data. Small samples can be used, for example, to draw conclusions about a very large population, within a clear and specified range of reliability.

Included herein are some subjects not normally included in statistical books, such as present value analysis and the internal rate of return, because, while these are widely used in the business world, they are often misunderstood by executives in the decision-making process.

I have also included a brief chapter on how to program a personal computer. This is not the great mystery that it is often portrayed to be. Computer languages are logical and straightforward, and some are very easy to understand. This little chapter will give you the fundamentals and teach you how to write some useful programs of your own.

I think you will find this an easy book to read. There are very few formulas to cope with, and throughout the book very simple and clear examples have been chosen to illustrate the various techniques.

You may prefer to read one chapter at a time, in a half-hour or so, and get a good grasp of that material before moving on to the next chapter.

And once you have completed the entire book, you may find it useful to wait a few weeks and then read it through again. In this way, your subconscious will have time to work, and anything that puzzled you at first may be entirely clear on this second reading.

There are exercises in the Appendix that can be used to test your grasp of the techniques explained in each chapter. And the book itself, of course, is a permanent reference that you can return to in the future to refresh your memory on any phase of the subject.

My guess is that you will find all of this to be exactly the opposite of dull. It is, in fact, a fascinating subject and the techniques can be great fun to work with, opening new doors to understanding that you never knew existed.

Knowledge is power, and that is what this book offers. I wish you well with it!

CONTENTS

1	Organizing Data	1
2	Frequency Distributions	9
3	Charts	17
4	Averages	36
5	Weighted Averages	43
6	Variability	47
7	Ratios and Percentages	53
8	Compounding	58
9	Present Value Analysis	62
10	Internal Rate of Return	67
11	Probability	71
12	Permutations and Combinations	78
13	Statistical Inference	83
14	Interpreting Survey Results	101
15	Sampling Techniques	108
16	Chi Square	114
17	Analysis of Variance	120
18	Regression and Correlation	125
19	Trend Analysis	136
20	Seasonal Variation	140
21	Compound Growth Rates	156
22	Index Numbers	162
23	Significant Figures	169
24	Statistics and Scientific Method	173

25	Personal Computers	183
	Appendix	193
	Index	223

BASIC BUSINESS
STATISTICS
FOR MANAGERS

CHAPTER 1
Organizing Data

The first step in any statistical analysis is to organize the basic data to be used. In some instances, indeed, this is enough to complete the entire analysis.

To illustrate this, assume that you are assistant to the sales manager and are asked to analyze the performance of each sales representative in the prior year. You ask the accounting department for the basic data, and they give you Table 1.1.

Here you see for each person, listed alphabetically, the assigned region, the percent of quota for the year, and the total sales in thousands of dollars. The data in this format are useful for checking individual performance, but otherwise have little analytical value.

The data are rearranged in Table 1.2 so that sales representatives are listed in rank order based on percent of quota. Now you see, at a glance, the relative performance of the sales group: the best, the worst, and others in between.

Next, you can arrange the data in rank order based on total sales, as shown in Table 1.3. Here again, you can quickly see how the sales representatives compare in this respect.

Finally, you can group the sales representatives by region, with the results shown in Table 1.4. Now you can see how the regions, and the sales representatives within each region, compare.

So with nothing more than this simple rearrangement of the data, you end up with several tables that provide a fairly complete analysis of how this sales group performed in the previous year.

The efficient way to do this, of course, is to put the basic data in a computer data base. The data can then be sorted and rearranged in a minute or two, and printed out accordingly, with a minimum of effort.

In any event, this simple process has converted a mass of confusing data into a rather sharp analytical tool that should be quite helpful to the sales manager in considering questions of compensation, allocation of accounts by salesperson and by region, and so on.

TABLE 1.1

Sales representative	Region	Percent of quota	Sales ($000)
Allen	5	115	1,741
Baker	5	108	2,915
Brown	4	107	1,811
Chesterman	3	92	2,176
Claytor	1	88	2,972
Dabney	6	135	1,889
Douglas	3	79	1,660
Erhart	6	113	1,615
Fulihan	6	122	1,417
Green	1	135	2,916
Hart	2	115	3,266
James	5	88	2,301
Lee	2	93	1,872
Long	1	167	2,667
Moore	6	108	1,971
Purvis	2	84	2,499
Roell	4	153	2,003
Shea	4	141	2,543
Smith	1	128	3,741
Taylor	3	89	3,114
Tower	4	93	1,432
Trice	2	147	1,741
Williams	5	103	2,215
Woolridge	3	97	2,975

The key lesson here is not to be afraid to rearrange the same data in several ways when each arrangement, in effect, answers a different question of substance. This is often the most efficient way to analyze a given subject.

COLLECTION OF DATA

In any statistical inquiry, there are two ways of gathering data. One is simply to accumulate any data that may be relevant. The other is

TABLE 1.2

Sales representative	Region	Sales ($000)	Percent of quota
Douglas	3	1,660	79
Purvis	2	2,499	84
James	5	2,301	88
Claytor	1	2,972	88
Taylor	3	3,114	89
Chesterman	3	2,176	92
Tower	4	1,432	93
Lee	2	1,872	93
Woolridge	3	2,975	97
Williams	5	2,215	103
Brown	4	1,811	107
Moore	6	1,971	108
Baker	5	2,915	108
Erhart	6	1,615	113
Hart	2	3,266	115
Allen	5	1,741	115
Fulihan	6	1,417	122
Smith	1	3,741	128
Green	1	2,916	135
Dabney	6	1,889	135
Shea	4	2,543	141
Trice	2	1,741	147
Roell	4	2,003	153
Long	1	2,667	167

to specify the precise questions that the inquiry should answer and to gather only those data that will answer these questions.

In analyzing most business problems the second approach is the preferable course to follow and yet, in practice, it is not uncommon to see examples of the first type in the form of voluminous reports with a great mass of data, often with no very clear relationship to the subject being analyzed.

I have seen many examples of this in the business world, sometimes in the form of a report written by consultants at a cost of hun-

TABLE 1.3

Sales representative	Region	Percent of quota	Sales ($000)
Fulihan	6	122	1,417
Tower	4	93	1,432
Erhart	6	113	1,615
Douglas	3	79	1,660
Trice	2	147	1,741
Allen	5	115	1,741
Brown	4	107	1,811
Lee	2	93	1,872
Dabney	6	135	1,889
Moore	6	108	1,971
Roell	4	153	2,003
Chesterman	3	92	2,176
Williams	5	103	2,215
James	5	88	2,301
Purvis	2	84	2,499
Shea	4	141	2,543
Long	1	167	2,667
Baker	5	108	2,915
Green	1	135	2,916
Claytor	1	88	2,972
Woolridge	3	97	2,975
Taylor	3	89	3,114
Hart	2	115	3,266
Smith	1	128	3,741

dreds of thousands of dollars. I must confess that my rather cynical reaction to this is a strong feeling that the primary purpose of the report is to impress the client with the great depth of research that has been conducted. In each such case, the difficulty becomes one of sorting through all the data to see if it has any real relevance to the problem at hand.

For instance, there was a recent news account of a chief executive who explained why he had decided to locate a major operations unit

TABLE 1.4

Sales representative	Region	Percent of quota	Sales ($000)
Long	1	167	2,667
Green	1	135	2,916
Claytor	1	88	2,972
Smith	1	128	3,741
Subtotal			12,296
Trice	2	147	1,741
Lee	2	93	1,872
Purvis	2	84	2,499
Hart	2	115	3,266
Subtotal			9,378
Douglas	3	79	1,660
Chesterman	3	92	2,176
Woolridge	3	97	2,975
Taylor	3	89	3,114
Subtotal			9,925
Tower	4	93	1,432
Brown	4	107	1,811
Roell	4	153	2,003
Shea	4	141	2,543
Subtotal			7,789
Allen	5	115	1,741
Williams	5	103	2,215
James	5	88	2,301
Baker	5	108	2,915
Subtotal			9,172
Fulihan	6	122	1,417
Erhart	6	113	1,615
Dabney	6	135	1,889
Moore	6	108	1,971
Subtotal			6,892
Total			55,452

in a certain city. He described about 50 criteria that were examined, in comparing various cities, as part of a comprehensive analysis of a great mass of statistical data. With total candor, he then concluded that, since operating costs appeared to be about the same in the several cities under consideration, he simply ignored all the other data and made the final decision on a purely subjective basis!

There is nothing wrong with a decision such as this, but it is clear that gathering all of this irrelevant statistical data served no useful purpose other than to create at least the initial illusion of a highly scientific approach to the problem. All this is no more than window-dressing, pure and simple.

It is also true that the shotgun approach to a statistical inquiry can stem from an unwillingness to do the hard mental work needed for developing the specific questions to be answered. It is often easier simply to accumulate data, without much concern for its real relevance. But this is almost certain to be a time-consuming and inefficient approach.

Here, by way of contrast, is an illustration of the rifle-shot approach. A good many years ago, after World War II, the state of Virginia needed extensive new school construction and I was asked to design a statistical study that would forecast the funds that would be required in the next ten years in each city and county in the state. This design took the form of specific questions that had to be answered in each case: current number of pupils by grade, which could be extrapolated year by year in the future; estimates of future population growth; classrooms in buildings already available; trends in construction costs; and so on and on.

Given these specific questions, it was possible to develop a systematic approach to arrive at the appropriate answers; and once these were available, they led directly to the desired conclusion as to total funds that would be needed by each local unit of government. This was a massive research project, but the design proved to be efficient and the entire study was completed very quickly. I was later told that the same design was adopted in its entirety by several other states with the same problem to solve.

Very few statistical projects in the business world are of the magnitude of this study, but the same principles apply. The key, in each case, is to start at the end of the study, and then work backward. What is the final result that you want to know, and what is needed, step by step, to get there? Whether the project be large or small, this type of rigorous analysis in the beginning will preclude the gathering of irrelevant data and produce a more efficient research design.

Assume, for example, that you are a bank officer charged with the responsibility of analyzing whether a branch bank should be established in a given location. How do you proceed with such an analysis?

The end result that you are looking for in this case is the estimated profit or loss from the branch operation each year for a certain number of years in the future. This is the return to the bank on whatever investment is required and clearly is a major determinant in whether or not the branch should be authorized.

In order to estimate profit you must first estimate revenue, and since this in turn is based primarily on volume of deposits, you must also have an estimate of that. Then you need to estimate the cost of running the branch.

To estimate deposits, you must first estimate total deposits generated within the trade area of the branch and then, taking into account the competitive branches serving the area, estimate your share of the market.

In all of this you will likely make use of a variety of data including internal records on other branch operations and the bank's historical experience in opening new branches. And you may well turn to outside sources for estimates of population growth, commercial development in the area, and so on.

Here, as often is the case in the business world, you are dealing with estimates and forecasts, and there is no precise way to assess the probability of error involved. About the only practical guide is to look at the relative accuracy of similar projections in the past for other branches that were established, where the same estimating system was used. And for this reason alone, it is most desirable to have a logical analytical format for conducting each study and rigorously adhere to it in each case.

SOURCES OF DATA

A staggering amount of statistical data is now being compiled and made available by a wide variety of public and private agencies. Among the public agencies, the U.S. Bureau of the Census and the U.S. Bureau of Labor Statistics are major providers of data.

8 BASIC BUSINESS STATISTICS FOR MANAGERS

In addition, many private companies have developed comprehensive data bases in many fields that can be accessed by computer for immediate answers to specific questions. And this resource is certain to grow in the years ahead.

In terms of internal data, the accounting departments of most companies have a great store of information, historical and otherwise, if the accountants can be persuaded to dig it up and make it available.

So the business executive today is very fortunate, in comparison with those of the past, in having this great wealth of data to utilize in the analysis of various business problems. The main difficulty is where to turn for the relevant data, and a good reference librarian can be most helpful in this respect.

SUMMARY

Many statistical inquiries in the business world require no more than a proper organization of the basic data. Very often the data can simply be rearranged in various ways to answer the different questions involved in the inquiry.

In general, the best way to conduct a statistical study is to start at the end, at the final result that is required, and then to work back step by step for the data needed to arrive at this final figure. This requires a rigorous analysis before the first piece of data is gathered and is referred to as the rifle-shot approach. It is different from the shotgun route, which is simply to accumulate a great deal of data and hope to find some solution therein. Although clumsy and clearly inefficient, the latter practice is all too common in the business world.

CHAPTER 2
Frequency Distributions

In their search for answers to various problems, executives are often presented with, and very often daunted by, statistical data in the raw that appear to be little more than a confusing mass of numbers. How do you deal with this situation?

As the preceding chapter illustrated, very often the first thing to consider is whether there is any way to reorganize the data in such a way as to bring some order out of chaos. And, as we have seen, this can sometimes be done by nothing more than putting the data in a simple array based on their rank order. But this system fails when a large number of data items are involved. For this we need an extension of the array concept, something called a *frequency distribution.*

Suppose, for example, that the sales manager of your company is interested in the productivity of the sales force in terms of calls per day. The basic data, taken from daily sales reports, are shown in Table 2.1, but no one can make any sense of the data in this format.

TABLE 2.1
Sales calls per day
(50 sales representatives over a period of 5 days)

3 5 3 3 4 4 4 6 4 5 5 3 3 3 4 5 3 2 2 5 3 4 6 4 3
3 3 5 4 3 2 3 2 5 1 1 4 5 4 2 4 5 4 5 4 4 4 5 2 2 3 1
3 4 2 3 5 2 5 4 2 6 4 4 5 2 3 4 3 4 5 5 4 4 2 5 3 2
2 1 1 3 3 2 4 2 4 5 6 4 3 3 2 3 6 2 5 2 5 4 5 3 4 4
1 4 5 4 5 4 1 3 4 4 6 3 1 3 6 2 5 2 2 4 5 4 3 5 5 3
4 4 5 2 2 3 6 4 5 1 3 4 3 2 4 4 4 5 5 4 5 4 3 3 5 5
3 3 2 5 2 4 3 4 2 6 4 6 6 3 5 3 3 3 3 4 5 5 3 5 3 3
3 3 3 5 6 3 2 5 2 4 4 2 3 6 3 4 3 6 6 3 2 5 5 4 3 1
2 2 4 2 3 4 3 2 5 1 4 4 2 3 3 5 2 4 3 3 4 4 4 4 2 3
5 6 5 5 4 3 1 5 2 6 3 3 2 3 3 5

The solution is simply to tally the frequency of days by calls per day, yielding this summary table:

Calls per day	No. of days	Percent
1	12	4.8
2	41	16.4
3	68	27.2
4	62	24.8
5	50	20.0
6	17	6.8
Total	250	100.0

From this it is evident at a glance that in a majority of cases sales representatives are making either three or four calls per day. In a smaller, but still significant, number of days there are as few as two calls or as many as five calls per day. Relatively infrequent is either one call or six calls per day.

To bring order out of chaos in this case, the only thing needed is to summarize the data in a simple frequency distribution, with the percentage column being added to aid in its interpretation.

It should be noted that it is very easy to compute the arithmetic average from this frequency distribution. Simply multiply the class average in each case by its frequency, add up these products, and divide by the frequency total. In this case multiply 1 by 12, 2 by 41, 3 by 68, and so on. The sum of these products is 898, and when this is divided by the frequency total of 250, you get 3.6 as the average number of calls per day.

The reason for this procedure is quite obvious. Normally, to get the arithmetic average you would add all the calls per day and divide by the total number of days. But clearly, you get the same result when you multiply calls per day in each class by the number of such days to arrive at the total. There clearly is no point in adding one 12 times or two 41 times when you can multiply 1 by 12 or 2 by 41 and get the same answer.

SELECTION OF CLASS INTERVALS

Now, let's turn to a more complicated example. Assume that you are considering a pension plan in a company plant and are interested in the age of its production workers. The basic data, in Table 2.2, is once again a confusing mass of numbers, and you want to reduce them to a more understandable form in a frequency distribution.

But here, unlike our first example, it is not feasible to designate classes by any single digit figure. Such a table, starting at age 20 and increasing a year at a time to age 70, would be almost as hard to decipher as the original data. To accomplish our purpose, we must select *class intervals* in which to tabulate the data. Two such selections are shown in the table: one with a class interval of ten years, the other with an interval of five years.

This selection of class interval is the key problem in creating any frequency distribution. There are no hard and fast rules on how this should be done. Clearly, the wider the interval, the greater the loss in detail. On the other hand, if the interval is too small, the table becomes too large and the real meaning of the data may well be obscured in a mass of detail. So this comes down essentially to a matter of judgment, based on the specific use to be made of the frequency distribution.

In the example given, there is obvious merit in class intervals ending in zero or five because we tend to think of these as key numbers. It would be rather awkward, for example, to have class intervals from 22 to 27, then 27 to 32, and so on.

A clear exception to this, however, is where there is some reason to think that data are clustered at various points. In such a case, it is desirable to fix the class intervals so that the clusters fall near the middle of each. The reason for this is that various computations made from a frequency distribution, such as the arithmetic average calculated in our first example, assume that the middle of the class represents the class as a whole.

An illustration of this is survey data in which people are asked to specify their age. Experience has shown that respondents tend to give ages ending with zero or five much more often than could possibly be true. In this case, if averages or other statistics are to be calculated from the frequency distribution, it would be preferable to have these numbers in the middle of the class intervals.

This exception would not apply, incidentally, to our age example where the data are presumably taken from accurate personnel records, free of any respondent reporting error.

You should always be very clear, and the table should clearly specify, where each class interval begins and ends. For example, if

TABLE 2.2
Age of production workers

30 63 34 37 42 49 51 43 54 56 31 35 38 32 43 56 27 24 62 37
52 67 50 33 38 39 55 47 39 24 31 22 55 49 61 50 21 47 54 40
56 43 47 61 24 34 31 51 26 33 63 23 53 42 26 45 41 62 29 41
34 52 53 62 44 49 22 53 31 26 30 28 24 48 51 65 52 28 39 27
64 25 58 46 58 36 47 50 41 59 49 58 46 32 42 49 67 35 36 68
23 55 21 22 50 55 48 37 58 56 29 42 40 59 24 22 34 67 42 65
38 43 33 25 52 53 50 66 66 47 57 48 35 31 55 55 35 37 23 59
48 39 42 40 67 34 60 33 32 38 35 48 63 61 37 55 27 31 38 38
30 60 68 33 53 40 52 21 31 36 46 39 70 36 25 61 58 50 34 48
23 36 47 30 60 40 44 22 36 36 63 50 37 32 52 40 40 42 23 28

From	To	Frequency
Under...	20	0
20	25	18
25	30	14
30	35	26
35	40	29
40	45	23
45	50	21
50	55	23
55	60	20
60	65	15
65	70	10
70	75	1
Over...	75	0
Total...		200

From	To	Frequency
Under...	20	0
20	30	32
30	40	55
40	50	44
50	60	43
60	70	25
Over...	70	0
Total...		200

you say "From 20 to 25," the assumption is that you mean "From 20 up to but exluding 25." If the data are carried to one decimal point, this same class interval would include everything from 20.0 up to and including 24.9. Again, it is important to have this clear in order to be certain that the data are tabulated correctly and also to be able to fix the true midpoint of each class for purposes of computation.

By the same token, open-ended class intervals such as "60 and over" should be avoided if possible. There clearly is no way to establish the midpoint of such a class.

It is preferable to have equal time intervals for all the classes, but with some data it may be necessary to change the interval as the table progresses. With tabulations based on family income, for example, it may be desirable to have wider intervals at the higher income levels. Whenever such a situation occurs, I would recommend a blank line at that point in the table to alert the reader to the fact that the interval is being changed.

COMPLEX ANALYSES

Thus far we have dealt with very simple examples, but the frequency distribution can be equally effective in summarizing a wide range of complex data that otherwise would defy analysis and understanding. A good illustration is Table 2.3, extracted from the *Media General Financial Weekly*. Here we see the return on equity—a key index of corporate efficiency—for all the major companies of the United States.

Since frequencies are shown in percentage form, you can compare, in consecutive columns, the companies listed on each of the major stock exchanges and their performance this year versus last year. A tremendous amount of information is summarized in this single compact table. Any chief financial officer, almost at a glance, can see where his or her own company falls in this total corporate spectrum.

From all of this it is evident that the frequency distribution is more than just an analytical aid. For many kinds of data, it is indeed an essential tool of analysis for which there is no real substitute.

FREQUENCY DISTRIBUTION CHARTS

Data in a frequency distribution can be plotted in a special bar chart called a *histogram*, as shown in Figure 2.1, where the class intervals

TABLE 2.3

Return on Equity

Stocks by Return on Equity Ranking*	Composite		NYSE		ASE		OTC	
	Now	52 Weeks Ago	Now	52 Weeks Ago	Now	52 Weeks Ago	Now	52 Weeks Ago
Top 5%	33.5	28.9	31.9	26.7	36.5	29.4	33.9	30.4
10%	24.5	22.7	23.7	21.2	26.7	22.9	24.6	24.0
15%	20.8	19.8	20.6	19.1	21.1	19.5	20.8	20.2
20%	18.4	18.2	18.3	17.8	18.7	18.3	18.4	18.5
25%	16.9	17.0	17.0	16.8	17.0	17.1	16.9	17.1
30%	15.9	16.1	16.1	16.2	15.9	15.5	15.8	16.1
35%	14.9	15.2	15.5	15.5	13.7	14.4	14.8	15.1
40%	14.0	14.4	14.7	14.8	12.8	13.4	13.8	14.3
45%	13.2	13.5	13.9	14.1	11.9	12.7	13.0	13.3
50%	12.3	12.7	13.4	13.5	11.2	11.9	12.1	12.5
55%	11.5	11.8	12.6	12.6	10.3	11.2	11.2	11.6
60%	10.7	11.1	11.7	11.7	9.4	10.4	10.3	10.9
65%	9.8	10.2	11.0	10.6	8.6	9.1	9.5	10.0
70%	8.9	9.1	10.2	9.9	7.7	8.2	8.5	8.9
75%	7.7	8.0	9.2	8.8	6.7	7.3	7.3	7.6
80%	6.7	6.9	8.0	7.8	5.8	6.5	6.3	6.5
85%	5.3	5.6	6.6	6.5	4.4	5.1	5.0	5.2
90%	3.8	4.0	4.6	4.8	3.3	3.7	3.7	3.5
95%	2.0	2.0	2.4	2.4	1.9	1.7	1.9	1.9
No. of Companies	3,752	3,407	1,165	1,163	488	478	2,099	1,766

Return on Equity Ranges (%)	Composite		NYSE		ASE		OTC	
	Now	52 Weeks Ago	Now	52 Weeks Ago	Now	52 Weeks Ago	Now	52 Weeks Ago
0.0 to 1.9	3.7%	3.7%	3.1%	3.3%	3.4%	4.2%	4.1%	3.8%
2.0 to 2.9	2.0	1.8	1.6	1.3	2.4	1.5	2.2	2.3
3.0 to 3.9	2.1	1.8	1.7	1.6	3.1	1.5	2.1	2.1
4.0 to 4.9	2.5	2.3	1.9	2.2	3.0	2.6	2.7	2.2
5.0 to 5.9	2.3	2.5	1.8	2.2	2.0	1.9	2.7	2.9
6.0 to 6.9	2.9	3.2	2.1	2.3	2.7	3.5	3.4	3.7
7.0 to 7.9	3.5	3.4	2.7	3.6	4.9	3.9	3.7	3.1
8.0 to 8.9	3.1	3.3	3.2	3.8	3.5	3.2	3.0	2.9
9.0 to 9.9	4.3	3.3	3.4	3.5	3.9	2.6	4.9	3.3
10.0 to 11.9	8.5	8.8	10.1	8.6	7.6	8.4	7.9	9.1
12.0 to 13.9	9.1	9.2	10.3	10.0	6.9	8.7	9.0	8.8
14.0 to 15.9	7.6	8.4	10.5	10.7	3.4	5.3	7.0	8.0
16.0 to 17.9	6.2	7.6	7.5	10.2	4.2	4.7	6.1	6.8
18.0 to 19.9	3.5	4.9	3.7	5.7	3.1	5.1	3.4	4.4
20.0 to 24.9	5.3	5.1	5.3	5.2	4.7	4.3	5.4	5.3
25.0 to 29.9	2.2	2.3	2.6	1.9	2.3	1.5	2.0	2.8
30.0 to 39.9	2.3	1.7	2.1	1.4	2.4	1.4	2.3	2.1
40.0 to 49.9	.9	.6	.8	.5	.4	.8	1.0	.6
50.0 And Over	1.6	.9	1.3	.5	2.0	1.0	1.7	1.2
Cannot Calculate	26.3	25.1	24.3	21.6	34.1	33.8	25.4	24.6
No. of Companies	5,092		1,538		740		2,814	

*Excludes companies with negative earnings and negative stockholder equity.

Source: Media General Financial Weekly. Reprinted with permission.

HISTOGRAM
SALARY DISTRIBUTION OF A GROUP OF EMPLOYEES

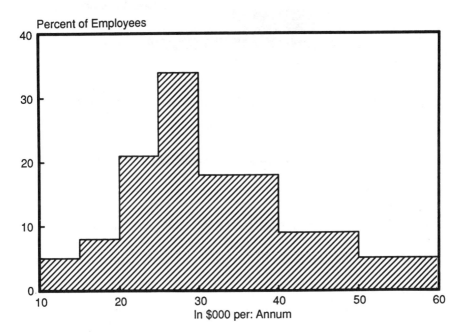

FIGURE 2.1

are plotted on the horizontal scale and the frequencies on the vertical scale. In such a chart, the class intervals may be equal or unequal, as long as they are plotted correctly. Where this is done, the area under the bar for any class is proportionate to its percentage of the total distribution.

Where the class is qualitative in nature, this area rule no longer applies. Here a bar chart can be used, as in Figure 2.2, where the length of the bar is proportionate to the size of the frequency involved. There is no basis for any variation in the width of the bars.

SUMMARY

The frequency distribution is an indispensable tool for reducing a large number of data items to a concise and understandable format.

16 BASIC BUSINESS STATISTICS FOR MANAGERS

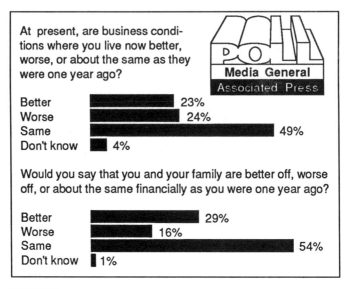

FIGURE 2.2

Source: Richmond Newspapers, Inc. Reprinted with permission.

Selection of class intervals to be used requires some thought and sound judgment. But beyond this, it is a very simple and straightforward process.

Putting the results in a simple histogram or bar chart often makes the underlying message clear at a glance and is especially useful in the executive environment where time is at a premium.

CHAPTER 3
Charts

Just as a picture may be better than a thousand words, a chart can often illuminate what a thousand figures would only obscure.

Charts really serve a dual purpose. They not only are quite effective as a presentation aid, but also often contribute to the analytical process itself by suggesting inferences and conclusions that may be far from self-evident in any inspection of the basic data itself.

Charts have special value in presentations to an executive group, where time is usually at a premium and the goal is to impart the maximum information in the minimum amount of time, because a good chart can condense a great deal of data and make it available almost at a glance.

At the same time, I would make one suggestion on this. Over the years, I have found that executives differ rather sharply on how they like to receive data. Some prefer text, others prefer charts, and still others like to see the actual data. The safe procedure, therefore, is to supply all three, even though some part of this may be attached as an appendix to the main report.

THREE KINDS OF CHARTS

Essentially, there are three kinds of charts, with a good many variants of each:

a. Pie chart
b. Bar chart
c. Line chart

The pie chart is a convenient way to show how a total is broken down into its component parts; each segment of the pie represents a corresponding part of the total. Figure 3.1 is a simple illustration.

18 BASIC BUSINESS STATISTICS FOR MANAGERS

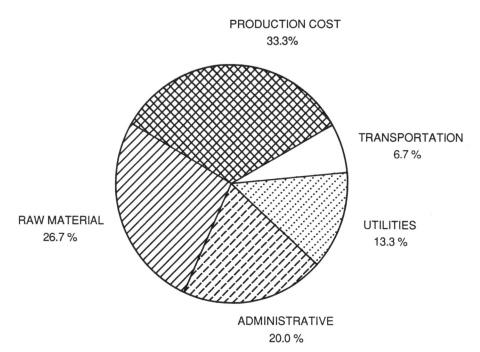

FIGURE 3.1

A bar chart also can be used to present the same data, as shown in Figure 3.2. It can run horizontally as well as vertically, as shown in Figure 3.3.

The choice among these three versions is really a matter of taste. In my own view, the relative size of each component is easier to grasp in a bar chart, but the pie chart is still a very compact way to illustrate the relationships involved.

One word of caution on bar charts. The base line should always be zero. To truncate the bottom portion of the chart by starting with some value other than zero can be a gross distortion of the data. This can be illustrated very simply. Suppose you want to represent the numbers four and two. If the bar charts use one instead of zero as

Charts **19**

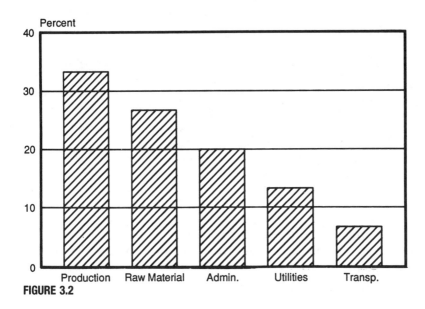

FIGURE 3.2

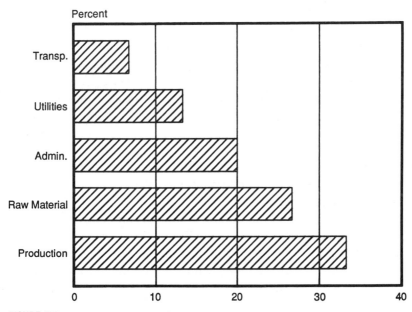

FIGURE 3.3

the base then the height of the bars will be as three is to one instead of the correct two to one, as shown in this diagram:

```
4 _
    |XXXXX|
3 _ |XXXXX|                                    _ 4
    |XXXXX|           |XXXXX|
2 _ |XXXXX|           |XXXXX|                  _ 3
    |XXXXX| |XXXXX|   |XXXXX|
1 _ |XXXXX| |XXXXX|   |XXXXX|                  _ 2
    |XXXXX| |XXXXX|   |XXXXX| |XXXXX|
0 _ |XXXXX| |XXXXX|   |XXXXX| |XXXXX|          _ 1

      A       B         A       B
```

Here the bars on the left start with zero as their base, those on the right with one as their base. The visual distortion in the latter is quite evident.

There are two principal versions of the line chart. The more common version has a vertical scale that is arithmetic. In another version, called a semilog chart, the vertical axis is scaled in logarithmic values.

The difference here is that a given slope on the arithmetic scale represents a uniform *amount* of increase, whereas on the logarithmic scale it represents a uniform *rate* of increase: a very fundamental difference indeed.

In other words, if a series is increasing by a constant amount such as 500 units per annum, the line drawn to represent the series will be a straight line in an arithmetic chart. If the series is growing by a constant rate such as 10 percent per annum, it will be a curved line upward in the arithmetic chart, but a straight line in a semilog chart.

The reason for this, of course, is that a straight line represents a constant slope and thus a constant amount of change; and a constant amount of change in logarithms translates into a constant rate of change in the original numbers.

Thus, if interest is primarily in rate rather than amount of change, the semilog chart is the one to use. In addition, it is also useful in comparing series of widely differing magnitudes where, on an arithmetic scale, the smaller series can be completely swamped by a scale drawn to include the larger series.

Incidentally, now that computers can draw charts quickly and with relative ease, computer-generated charts will undoubtedly be used more often in the business world. While this computer output tends to be something less than elegant, it is nevertheless quite adequate for typical business use.

EMBELLISHMENT OF CHARTS

For more formal presentations, charts can be embellished in a variety of ways including the use of color, three-dimensional representation, and the like. It is evident that modern chart makers are becoming ever more ingenious in producing effective graphic displays.

With three-dimensional charts, however, be on guard against one thing. If areas such as blocks or circles are shown, the area, and not its dimensions, should be proportionate to the number represented. If two squares, for example, represent the numbers 10 and 20, it is quite improper to draw the sides of the squares in the ratio of 1 to 2. In such a case, the eye sees the area, which is in the ratio of 1 to 4, and thus such a chart is grossly misleading.

This simple diagram illustrates the point. The dimensions of A are doubled in B, but the area is increased fourfold.

A   B

SELECTED CHARTS OF THE CONFERENCE BOARD

The best way to learn more about charts is to keep your eye open for good ones, and they can be seen on every hand. In my view, some of the best chart work in recent years has appeared in various publications of The Conference Board, and it may be useful to look at a few of these. Some were published in full color, and while much of their impact is lost in black and white reproduction, they are still excellent illustrations of various chart forms. (See Figures 3.4-3.16 at the end of this chapter for examples.)

SUMMARY

The three basic kinds of charts are pie, bar and line. Whether they appear in a simple or an embellished format, they can be highly effective, both in the analysis and in the presentation of statistical data. Their fundamentals are simple and easy to understand, and all executives can benefit by learning how to use them properly and effectively not only in their own work, but also in judging the work of others.

Postwar Trend in U.S. Productivity Growth
Average Annual Percent Change

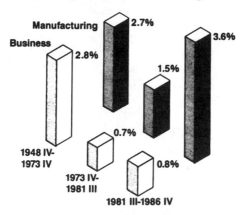

U.S. Productivity and the Business Cycle

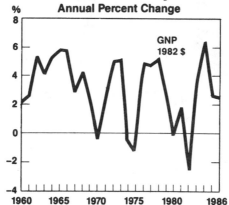

FIGURE 3.4

Three-dimensional bar charts combine with an arithmetic chart to depict this series.

Source: U.S. Department of Labor; the Conference Board.

24 BASIC BUSINESS STATISTICS FOR MANAGERS

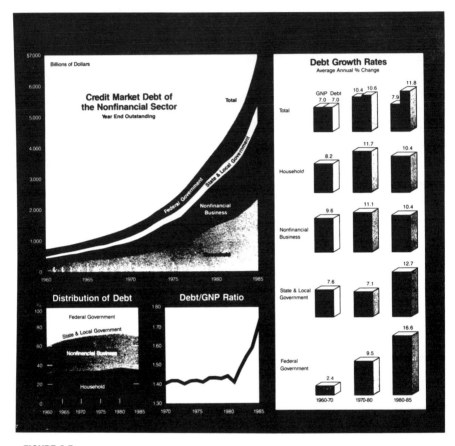

FIGURE 3.5
Here we have band charts to show components over time, on both an actual dollar and percentage basis, plus a bar chart and a line diagram. A complicated story, told almost at a glance.

Source: U.S. Department of Labor; The Conference Board.

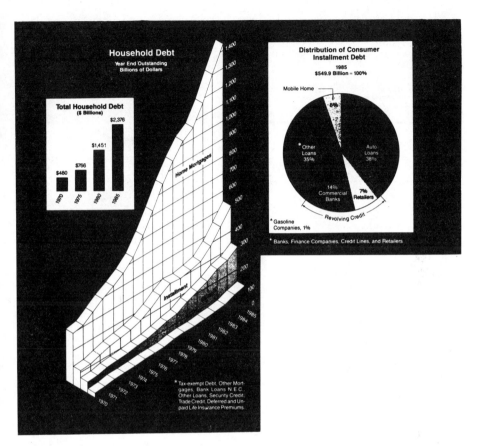

FIGURE 3.6
An interesting three-dimensional line diagram. The bar chart and pie diagram are useful supplements.
Source: U.S. Department of Labor; The Conference Board.

Cash Flow and Capital Spending

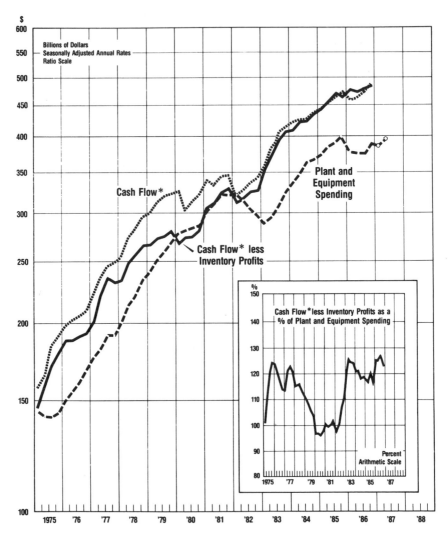

* Profits after tax plus charged depreciation (excluding adjustment for under-depreciation).
○ Anticipated
Sources: U.S. Department of Labor; The Conference Board

FIGURE 3.7
Here we have a semilog chart combined with a line diagram that pinpoints a key ratio. The two are an effective combination.

Charts **27**

International Asset Holdings

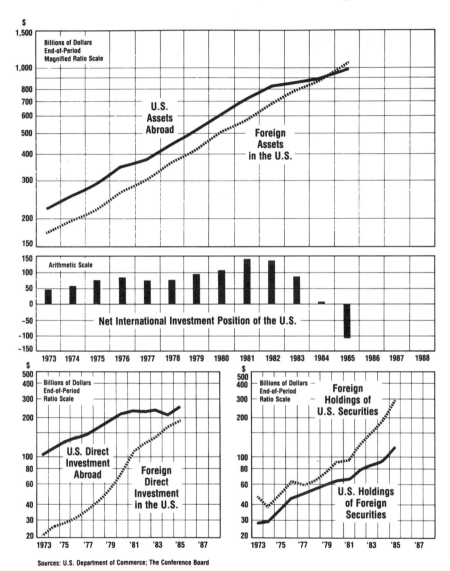

Sources: U.S. Department of Commerce; The Conference Board

FIGURE 3.8
Three semilog charts and a bar chart highlight the key trends here.

Wage Indexes

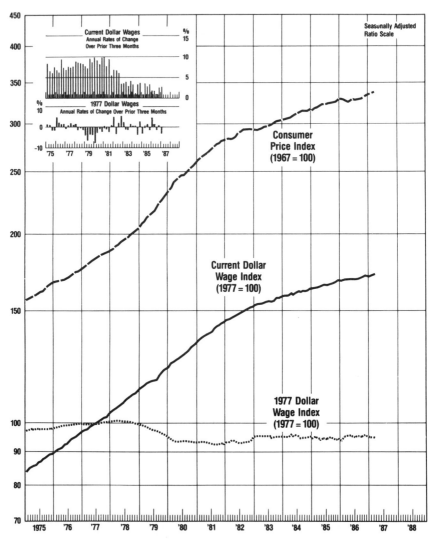

FIGURE 3.9
The semilog chart is a clear portrayal of the trends in the data, which the bar charts supplement with data on annual rates of change on both a current and constant dollar basis.

Growth of Civilian Employment

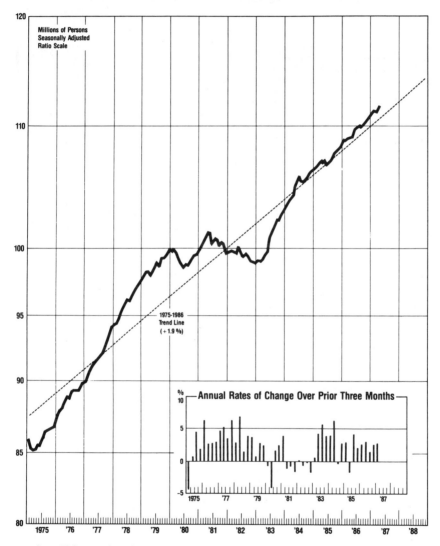

FIGURE 3.10

The trend line drawn in this semilog chart shows a constant rate of change throughout the period, against which the actual data may be compared.

30 BASIC BUSINESS STATISTICS FOR MANAGERS

The Growth of Debt and Its Relation to Income Flows

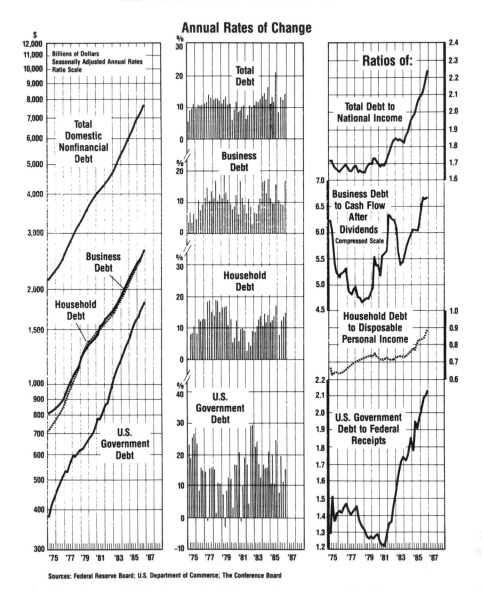

FIGURE 3.11

Here we have the combination of a semilog chart, bar charts to show annual rates of change, and arithmetic charts showing key ratios: a tremendous amount of information in a very compact space.

Three Levels of Goods Prices

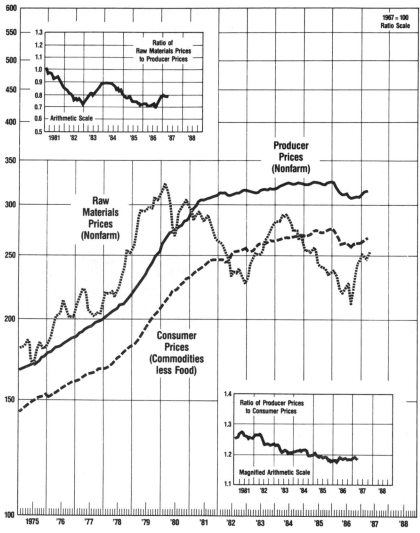

FIGURE 3.12
Note here the use of arithmetic and semilog charts, with a magnified arithmetic scales on the lower chart dealing with ratios. It is an excellent practice, as in this case, to cite the difference in scale for the benefit of the reader.

32 BASIC BUSINESS STATISTICS FOR MANAGERS

The Federal Debt

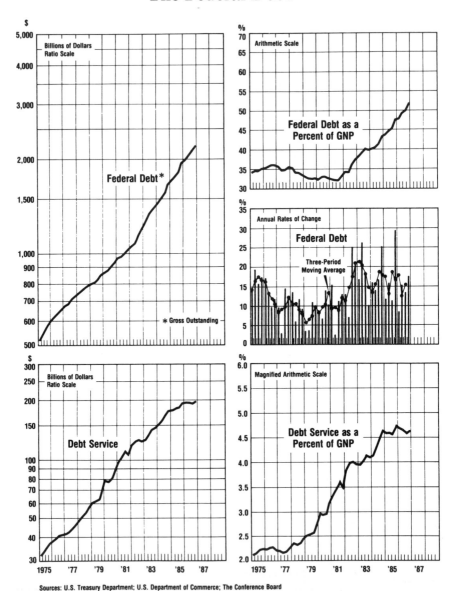

Sources: U.S. Treasury Department; U.S. Department of Commerce; The Conference Board

FIGURE 3.13
Again, a combination of semilog and arithmetic scale charts plus a bar chart on annual rates of change. All together they tell the story very clearly.

The Government Sector

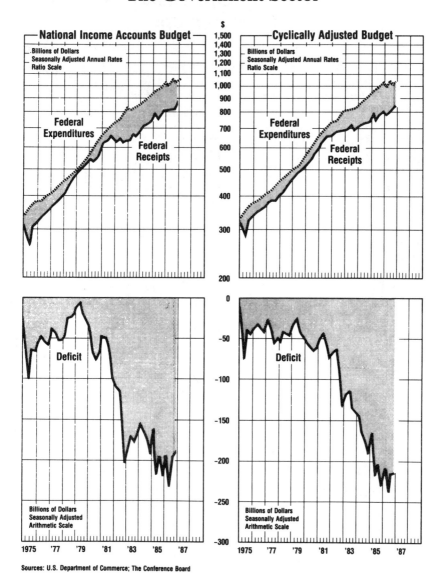

FIGURE 3.14
Here the federal deficit is clearly depicted both as the gap between the receipts and expenditure lines and in the deficit bands on the lower arithmetic charts.

Wages, Productivity, and Costs: Business Sector

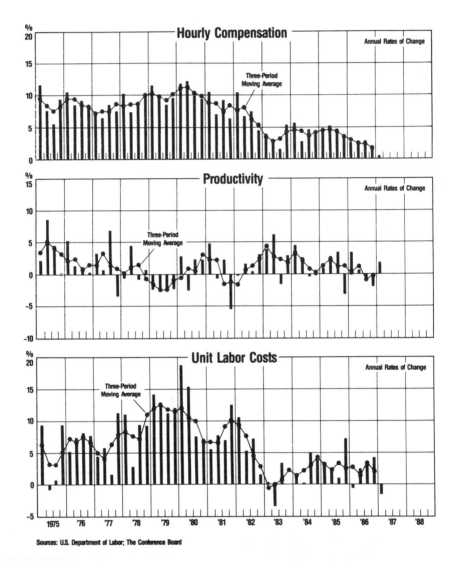

Sources: U.S. Department of Labor; The Conference Board

FIGURE 3.15
Here bar charts clearly show annual rates of change and also contain a line showing the three-month moving average in each case.

Determinants of Plant and Equipment Outlay

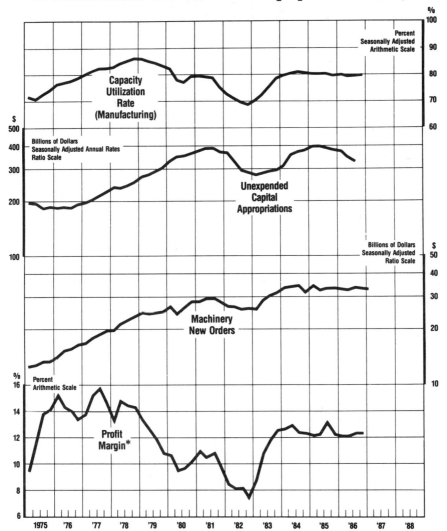

FIGURE 3.16
Another excellent use of semilog and arithmetic charts to pinpoint key trends in the data.

CHAPTER 4
Averages

The business executive must make judgments and many decisions based on averages of one kind or another. Some typical examples:

How do salaries in one division compare with those in other divisions of the company and with those of competitors?

What was the average performance of the sales force last year, and how does each member compare with this average?

How does the company's return on equity compare with the industry average and with the average for all corporations?

And so on. In comparing various sets of numbers, the use of an average as a summary measure of each is essential. It is also viewed as a routine and commonplace procedure, and I doubt that many executives have ever given any thought to what averages really mean.

ARITHMETIC MEAN

Everyone learns at an early age to compute an average. You simply add up a series of numbers and divide by the number of items. What more is there to be said on the subject?

Well, in the first place, I have only described one kind of average—what statisticians call the *arithmetic mean*—and there are other kinds you should know about, each of which has its own merits.

Many years ago, when I taught an evening class in statistics, my students would receive this statement with obvious scepticism. They had managed quite well over the years with what I was now calling an arithmetic mean, and the thought of anything beyond this seemed pretty foolish.

So I would put a little problem to them, as follows: a freight train going from one city to another went 20 mph for the first half of the trip, and 30 mph the second half. What is the average rate of speed?

Initial scepticism on the part of the students now turned into dark suspicion. The obvious answer of 25 mph must be wrong, but if so, why is it wrong, and what is the correct answer? Work it out in the following table:

Half	Miles	Hours	MPH
1st	60	3	20
2nd	60	2	30
Total	120	5	24

You see that if it takes five hours to go 120 miles, then clearly the average mph is 24 and not 25. (In this example, any distance between cities will produce the same end result.)

So, quite clearly, the old and familiar arithmetic mean doesn't work in this situation. Something else is needed to arrive at the correct answer. But before going on to that, it might be useful to consider a related trick question:

If the train goes 20 mph the first half of the trip, how fast must it go the second half to average 40 mph for the entire trip? The answer is that no speed in the second half, no matter how great, can accomplish this feat!

You can deduce this, almost at a glance, from the arithmetic of our first example:

Half	Miles	Hours	MPH
1st	60	3	20
2nd	60	0?	?
Total	120	3?	40?

With three hours already consumed in the first half of the trip, zero hours would be required for the second half in order to average 40 mph; and with any time allowed for the second half, no matter how infintesimally small, the average for the entire trip must be less than 40 mph.

As simple as this may look with the arithmetic in front of you, it is by no means self-evident to others. If your associates like a friendly wager now and again, you might win a few dollars with this one.

HARMONIC MEAN

But enough of fun and games, and back to the basic example. The arithmetic mean will not work in this case. Is there some other type of average that will work? Yes, there is such an average, and it is called the *harmonic mean*.

To find the harmonic mean of a series of numbers you simply compute the reciprocal of each (which, of course, is one divided by the number), add these, divide by the number of items, and then determine the reciprocal of that result. In our example, it would work this way:

Half	MPH	Reciprocal
1st	20	.05000
2nd	30	.03333
Sum		.08333
Divide by 2....		.04167
Reciprocal......		24.00000

And you get 24 mph as the harmonic mean, which you know by now is the correct answer to this particular problem. And, indeed, the harmonic is the only mean that provides the right answer when dealing with data of this kind.

As another example, a manufacturer buys $1,000 worth of a given supply item at $1.00 per unit and a month later buys another $1,000 worth at $2.00 per unit. What is the average cost? The arithmetic mean of $1.50 is incorrect. The proper answer is the harmonic mean of $1.33.

GEOMETRIC MEAN

Another average you should know about is called the *geometric mean*. To get the geometric mean of "N" numbers, you multiply by each number, and then take the Nth root of the product.

For example, to determine the geometric mean of 10 and 40, multiply the two numbers to get 400, and then take the second (or square) root of this product to get 20 as the appropriate mean.

For more than just a few numbers, logarithms should be used to make the calculation. You may recall that if you add the logs of numbers, you get the log of their product, and if you divide a log by N, you get the log of the Nth root of the number.

To see how this works, take 10, 20, and 30 as the three numbers, and calculate their geometric mean:

Number	Logarithm
10	2.3026
20	2.9957
30	3.4012

Sum................ 8.6995
Divide by 3..... 2.8998
Antilog............18.1712

Thus you find that the geometric mean of 10, 20, and 30 is 18.1712. You would have obtained the same answer, of course, by multiplying 10 by 20 by 30 to get 6000, and then taking the third root of that product.

It is no longer necessary to look up logarithms in a table. They are obtained with one keystroke on a little hand calculator; and the antilog is calculated just as easily. We have used natural logs in this example (which have 2.7182 as their base), because they have some advantage over common logarithms (on the base of 10) and are just as easy to deal with.

The geometric mean is the appropriate measure to use when dealing with ratios or index numbers and the primary interest is in proportional rather than amount of change from one period to another. For example, assume we have price indexes on two commodities as follows:

Mean	Commodity	Period 1	Period 2
	A	100	200
	B	100	50
Arithmetic...........................		100	125
Geometric...........................		100	100

The arithmetic mean shows an increase of 25 percent in the average price of these two commodities, whereas the geometric mean shows no change at all. This clearly demonstrates the difference between the two.

The arithmetic mean reflects the amount of change, with the increase of 100 points in A versus a decrease of 50 points in B resulting in an average increase of 25 points. The geometric mean, by contrast, gives as much weight to the halving of B as it does the doubling of A, and thus shows no net change at all for the two.

Which result is right and which is wrong? The answer is that they are both correct, but they answer different questions. If you are interested in amount of change, the arithmetic mean is appropriate. But if your concern is with proportional change, then the geometric mean gives you the answer you are looking for.

One use of the geometric mean, for example, has to do with the movement of stock market prices. Most stock indexes weight these price changes by market value of each stock, and large companies tend to dominate the indexes accordingly. The result can be quite different from an index based on the geometric mean, which gives equal weight to the proportional change in each stock.

Here is still another use for the geometric mean. Suppose you have a series with an increase over the preceding year of 10 percent the first year, 10 percent the second, and 40 percent the third. What is the average rate of increase per annum?

If you take the arithmetic mean of these three percentages, you get 20 as the answer, and this is wrong. The right procedure is to put these percentages in ratio form: 1.10, 1.10, and 1.40; and then find their geometric mean. This is 1.192, which represents an annual increase of 19.2 percent, and that is the correct answer.

MOST TYPICAL MEAN

The merits of still another type of average can be illustrated with these data on the number of television sets per household in a major market:

No. of sets	Percent of households
0	2
1	52
2	33
3	11
4	2
Total	100

To get the arithmetic mean, you multiply the sets per household by the percent of households, sum these products, and divide by 100: the answer is 1.6. This is a useful figure, but it leaves something to be desired if we wish to describe the typical household, which can no more have 1.6 television sets than it can have 1.6 children.

To describe the typical family we need what is called the mode: the value that occurs with more frequency than any other. In the data shown above, the mode is 1 set per household, and that is a better description of the typical household than any other value we could select in this situation.

It should be noted that these two means do not contradict, but rather supplement, each other. Each has its own story to tell. The mode tells us that the typical household has one television set, and the arithmetic mean tells us that since a good many households have more than one, the total number of television sets is 1.6 times the number of households.

MEDIAN

This brings us to the fifth and last of averages in common use. Consider these data on salaries in a small group in a company:

Employee	Salary
A	$ 15,000
B	16,000
C	18,000
D	20,000
E	100,000

What we have here is a department head and four assistants. What kind of average would best describe their salary distribution?

The arithmetic mean is $33,800, but it is clear that this is a very atypical figure in terms of any member of this group. It has the mathematical quality of being the number that, when multiplied by five, will equal the total payroll of the group. But beyond this, it tells us very little about the individual members of the group.

For this we must turn to the *median*, which is the middle value when the salaries are ranked from low to high. In this case, the median is the $18,000 salary of Employee C, which is much more representative of individual salaries in the group than the arithmetic average.

The median always has this precise meaning: that one-half of the items in the given distribution have a higher value, and one-half have a lower value, than the median. It is always the middle value, splitting the distribution in half. If there are an even number of items in the series, then the arithmetic mean of the middle pair is the median.

In the real world we must often deal with data, like this salary distribution, where a relatively small number of extreme values can so distort the arithmetic mean as to make it quite atypical. In these cases, the median is far more representative of the group.

But once again, there is no harm in showing both the arithmetic mean and the median in situations of this kind. The greater the deviation between the two, the greater the distortion created by the high-value members in the group.

This distortion can be found in many kinds of data. In comparing the performance of a given pension fund with that of other funds, for example, it is not unusual to find a few extreme values that will have a significant effect on the arithmetic mean. As a result, the median for the other funds would normally be preferred for this type of comparison.

SUMMARY

The attempt to summarize a series of numbers, and sometimes a very large series, with a single number is a formidable task, and it is not surprising that the effort is more complex than might first appear.

Statisticians refer to averages as *measures of central tendency*, and we have seen that there are five such measures in common use, each with its own merits and special application. The fact that they usually differ does not mean that one is right and the others wrong, but simply that each is designed to answer a specific question, and the questions differ.

Executives will rarely encounter some of these averages, but they should at least know of their existence and be aware of the reasons for their use. And if you in particular have absorbed the lesson of this chapter, you will feel much more comfortable in dealing with any problem in which averaging is involved.

CHAPTER 5
Weighted Averages

You may be a bit intimidated by the term *weighted average*. It has a mysterious sound that suggests all kinds of hidden complications. Happily, this is not true. It actually is a very simple concept, easy to understand and to use.

Consider the following situation. Your company has two plants making the same product. Last year, the cost per unit produced was as follows:

Plant	Per unit cost
A	$120
B	150

Now, the questions is, what is the average cost for both plants combined?

You might decide simply to get the arithmetic average of 120 and 150, and thus derive $135 as your answer. But the only way this could be correct would be for each plant to produce the same number of units. Assume, however, that Plant A produced 1,000 units and Plant B produced 2,000 units. Now, how do you solve the problem?

The necessary data now being at hand, the straightforward approach is to calculate the total cost of production in each plant by multiplying the unit cost by number of units in each case. This produces:

Plant	Total cost
A	$120,000
B	300,000
Total	420,000

Now, divide $420,000 by the 3,000 units produced by both plants, and get $140 as the average for both plants combined. This is the weighted average, with unit cost in each plant multiplied (or weighted) by the number of units produced by each. No more complicated than that!

It should be observed that the first answer of $135 is called an *unweighted average*. More correctly, it is a weighted average that assumes equal weight for each of the component items, a situation that clearly does not prevail in this case.

Also, it should be noted that it is not necessary to use the actual number of units produced as the weights in this case. Any other set of numbers could have been used, as long as they were in the same proportion to each other as the number of units. For example, you could have used a weight of 1 for plant A and a weight of 2 for plant B and gotten exactly the same result.

Clearly, the weighting process itself is very simple, and in practice the only difficulty is in selecting the proper weights to use. The best way to do this is to visualize what the product means when a given average is multiplied by a given weight, and see if this passes the test of logic.

In this example, the product represents the total cost of production in each plant, and it is evident that this is the proper concept to use in calculating the average unit cost for both plants combined.

If, for example, you were dealing with the average wage in each of the two plants, the logical weight to be applied would be the number of employees in each case. When average wage is multiplied by employees, the result is total payroll; and when this is aggregated for the two plants and divided by total employees, the result is clearly a proper average for the combined operation.

Similarly, if you are measuring the cost of a given household budget and you know, for example, the price of milk per quart, then this should be weighted by the quarts of milk in the budget. Or, if you are tracking stock prices, you can weight the price of each stock by its shares outstanding to get the total market value of all such shares. The key question, always, is whether the weighted product meets the test of logic.

CHANGE IN MIX

The change in weighted averages from one time period to another can sometimes appear to be very puzzling indeed. Let's expand our example to include two such periods, as follows:

| | **Unit cost** | | |
Plant	Year 1	Year 2	Change
A	$120	$125	+5
B	150	155	+5
Avg.	140	135	−5

At first glance, this looks quite impossible. How can unit cost go up in each plant and yet go down when the two are combined?

The answer, of course, is that a weighted average is always the product of two things: the individual average and its weight. Neither is more important than the other in producing the final result. And when you look at the result for one time period versus another, you must remember that changes can occur in the weights as well as in the individual averages involved.

In the table above, we assumed 1,000 units in plant A and 2,000 units in plant B in the first period, which was reversed with 2,000 units in plant A and 1,000 units in plant B in the second period. This produced a total cost of $420,000 in Year 1 and $405,000 in Year 2, for a weighted average unit cost of $140 and $135 respectively and a decline of $5 in unit cost from one period to the next.

And thus the apparent paradox disappears, once we recognize the mechanism involved in computing the weighted average in each case. But the example does make it clear that you should always be fully alert to the fact that a change in weight can materially affect the movement of weighted averages over time.

This is normally referred to as a change in *mix*, and sometimes it is useful to measure the effect of this change. To do this in our example, assume that production in each plant was the same in Year 2 as in Year 1. In this case, the total cost of production in both plants would have been $420,000 in Year 1 and $435,000 in Year 2, for an average unit cost of $140 in Year 1 and $145 in Year 2.

Thus, with no change in weights, unit cost would have increased by $5 from one year to the next. The actual change was a decline of $5. Hence, the change in mix had the effect of reducing unit cost by the difference between these two, or $10. This, of course, is simply another way of saying that while unit cost was down $5 from one year to the next, it would have been up $5 if the production in each plant, or the mix, had remained the same.

The example used here is both simplified and somewhat exaggerated to make the key point involved, but it does illustrate a useful technique in dealing with actual data in the real world. It is often very helpful to recalculate weighted averages with no change in weights in order to measure the effect of a changing mix.

It should also be noted that there other ways to recalculate the data with different weights. For example, unit production in the second year or the average for the two years might be chosen in recomputing the weighted averages. Different answers will be produced in each case, but in effect different questions are being asked, and it is appropriate that the answers should vary accordingly.

SUMMARY

In order to combine individual averages of component units into a consolidated average for the entire group, appropriate weights should be applied to each. Where this is not done, the implicit assumption is that each individual average should be given equal weight, and the user should be entirely clear on the fact that this assumption has been made.

The process of weighting is simple arithmetic and presents no special problems, but one must be quite certain that the proper weights have been selected. The test here is whether the product of the individual average times the weight is a logical quantity for the purpose at hand.

In the analysis of changes in weighted averages over time, it must be remembered that this is the compound result of changes in weights as well as in the individual averages involved. The effect of this change in mix can be measured precisely, and it is often useful to do so.

CHAPTER 6
Variability

For the executive, as for everyone else, variability in almost every circumstance is a fact of life and must be dealt with. But in terms of numbers, there are laws that govern variability, and it is important to know how they work.

In comparing one group with others, it is usually convenient, if not essential, to summarize each group in terms of some appropriate average. But while this a powerful analytical device, it only tells half the story. The other half has to do with the scatter around the average, giving rise to the worn joke about drowning in a river with an average depth of only a few inches.

To make this more tangible, consider the salaries of two sets of individuals:

A	B
$40,000	$20,000
40,000	40,000
40,000	60,000

Both groups are the same in terms of average salary, but clearly quite different in the variation around that average. So the question now becomes: is there some way, perhaps in a single number, to summarize this variation in comparing the different groups?

One obvious answer is to use the range in the series from the minimum to the maximum value. In the two series above, this would be zero for Group A and $40,000 for Group B.

Although the range is sometimes used, especially in comparing salary and compensation data, it has severe limitations. Being based on only two figures, the top and bottom of a series, it reveals nothing about the distribution in between. And beyond this, the range tends to increase with the size of the sample, and thus is an unstable value.

There are other positional measures of variation that may be derived from a series ranked from low to high. Decile values, for example, divide such a series into ten parts; quartile values into four parts; and percentile values into 100 parts. As an example, look at the price change of more than 4,000 common stocks in the 52 weeks prior to my writing this. The quartile values are as follows:

Quartile	Percent
1	−17.5
2	7.2
3	30.8

And you see that one-fourth dropped more than 17.5 percent, another one-fourth increased more than 30.8 percent, one-half increased more than 7.2 percent, and another half less than that.

One-half the difference between the first and third quartile, or 24.2 in this case, is called the semi-interquartile range, and approximately one-half of the stocks are within the range of the median (which, of course, is the second quartile) plus or minus this 24.2 range.

Similar inferences can be made from the decile values, which are as follows:

Decile	Percent
1	−43.0
2	−25.1
3	−12.0
4	− 0.8
5	7.2
6	15.5
7	24.8
8	37.9
9	58.3

This shows that one-tenth of the stocks lost more than 43.0 percent in value and one-tenth gained more than 58.3 percent. The difference between the two, or 101.3 percentage points, is the decile range of the series, and includes four-fifths of all the stocks.

The same process is involved with percentiles. Here are some selected values from the stock data series:

Percentile	Percent
1	− 82.0
5	− 58.8
95	82.0
99	140.0

Here you see that 1.0 percent of our stocks lost more than 82.0 percent in value in the preceding year, and 5.0 percent lost more than 58.8 percent, while another 5.0 percent gained more than 82.0 in value, and 1.0 percent gained more than 140 percent: a graphic illustration of the potential for both profit and loss in the great roulette wheel on Wall Street!

These are useful numbers in the analysis of data, but they all suffer from the disadvantage of not being amenable to algebraic treatment. Beyond this, they are not as compact as we would like. Ideally, just as one average can summarize the central tendency of a series, it would be helpful to have a single number to summarize its variation.

AVERAGE DEVIATION

To pursue this, take the following simple set of numbers, with the deviation of each from the arithmetic mean:

	X	d	
	1	−3	
	2	−2	
	3	−1	
	6	2	
	8	4	
Sum.......	20	12	(ignoring signs)
Average.......	4	2.4	

It is a characteristic of the arithmetic mean that all deviations from such a mean will always sum to zero, which is a useful quality but not very helpful in the search for some measure of variability to compare one group with others. The only way to overcome this is to ignore the signs of the deviations.

In the case above, for example, if you treat all deviations as positive, they sum to 12, and if you divide this by the number of items, you get 2.4 as the average deviation. This is a straightforward process, and it yields the desired single-figure measure of variability. It also has the merit, unlike positional measures, of using all of the data in the series, with each item contributing to the average.

The chore of computation can be eased somewhat by the following formula, where A is the sum of items above the average, B the sum of those below, a the number above, b the number below, and M is the arithmetic mean:

$$\text{Sum}(d) = A - B - (a - b)M = 14 - 6 - (2 - 3)4 = 12$$

While the average deviation seems attractive as a solution to the problem, it does have one major disadvantage: just as in the case of positional measures of dispersion, it is not subject to algebraic treatment. The laws of algebra are not very flexible when it comes to signs, and there is no way to specify in a formula that negative values should be treated as though they were positive.

Thus, although it has the merit of simplicity, the average deviation is rarely used except for a few specialized tasks such as measuring the spread between real estate assessments and actual values. Statisticians turn instead to another measure that is not only algebraically correct, but also highly useful in deriving general laws of statistical inference.

STANDARD DEVIATION

This measure is called the *standard deviation*. It also is based on the deviation of each item from its arithmetic mean, but it uses a different averaging process for these deviations: a *root mean square average*.

As the name suggests, to derive this average you first square all the deviations, divide the sum of such squares by the number of items, and then take a square root of the result in order to return to the original unit of measure.

With X as the value of each item, M the arithmetic mean, and N the number of items, the formula for the standard deviation is:

$$S = \sqrt{(\Sigma X - M)^2/N}$$

Variability 51

This may look formidable, but it really is quite simple. Take the prior example to see how it works:

	X	d	d²	X²
	1	−3	9	1
	2	−2	4	4
	3	−1	1	9
	6	2	4	36
	8	4	16	64
Sum	20	0	34	114

Now, the standard deviation is simply the square root of the sum of d^2/N. In this case it is the square root of 34/5, or 6.8, and equals 2.61. Thus the problem of the average deviation, which must ignore signs, is eliminated in this measure of variation.

And beyond this, oddly enough, the standard deviation is really easier to compute than the average deviation because of the following short-cut method:

$$\text{Sum } (d^2) = \text{Sum } (X^2) - (\text{Sum}X)^2/N$$

In this example these values work out to be:

$$\text{Sum } (d^2) = 144 - 20^2/5 = 114 - 80 = 34$$

This is much easier than finding the deviation of each item from its mean, which may involve decimal values, and then squaring and adding the squares of each. By contrast with this tedious procedure, the shortcut formula is simple with a hand calculator and child's play with a computer.

There is one modification in the formula for the standard deviation when working with samples. In this case, the divisor in the root mean square average should be N-1 (which is the degrees of freedom in the sample with the average itself taking up one degree) instead of N. This makes little difference in large samples, but can be significant when the sample is relatively small.

COEFFICIENT OF VARIATION

Finally, to compare the variability in one series with that in others, there is the *coefficient of variation*, which is simply the standard deviation as a percentage of the arithmetic mean. This is particularly useful when the means differ considerably in value.

Suppose, for example, there are two plants that have the following daily production data over a period of one month:

	A	B
Mean...............	200	400
Stardard deviation.	20	32
Percent............	10	8

It would be wrong to conclude from these data that daily production in Plant B is more variable than that in Plant A because it has the greater standard deviation. When account is taken of the size of production in each plant, in the coefficient of variation, it is clear that Plant A is the more variable of the two.

SUMMARY

In conclusion, therefore, it is evident that you can summarize a great deal of data with just two figures: the arithmetic mean and the standard deviation, the former depicting the average of the data and the latter the variation around that average.

And, as a later chapter will show, these two powerful numbers do far more than summarize a set of data. In a great many cases, it is possible—using nothing but these two numbers—to closely approximate the shape and character of the entire distribution.

This quality, in turn, lays the foundation for various laws of statistical inference, which are far-reaching in scope and have many important applications in everyday life.

CHAPTER 7
Ratios and Percentages

Your task as a business executive is to make sense out of the numbers that you work with, and this usually requires placing them in a proper context. Typically, one number by itself has limited meaning, but when related to another number, it can be quite significant. This can be accomplished through *ratio* analysis.

To get the ratio of one number to a base number you simply divide by the latter. The ratio of ten to two, for example, is five. This is the same as five to one: for every one of the base number there are five of the given number.

It is often convenient to change the base. Multiply the ratio by one-hundred and it becomes a percentage: the ratio of ten to two is five-hundred percent.

All this is purely a matter of convenience. Generally it is preferable to modify the base in such a way as to get whole numbers, or numbers with no more than one decimal figure, in the resulting ratio. For instance, an accident rate per working hour might be something in the order of .00053. While this ratio may be quite correct, it certainly is not very readable. It is easier to grasp when expressed as 53 accidents per 100,000 working hours, or 5.3 accidents per 10,000 hours. All of these, of course, mean exactly the same thing.

To illustrate the point that a single number by itself has limited meaning, suppose that your company has stockholder's equity, or net worth, of $23,902,000. This reveals something about the size of the company but not much else. Now, assume that the income of the company is $4,117,000. Dividing this by the total of net worth gives a return on equity of 17.2 percent.

Now, while neither the net worth figure nor the income figure by itself can tell but so much, the ratio of the two is highly significant. It is a key index of corporate performance and can readily be compared with similar ratios for other companies as a measure of relative performance.

This is one of the great advantages of ratio analysis. Ratios are abstract numbers and can be compared directly with other ratios of the same kind, irrespective of how the original numbers may vary in size.

Take the accident rate example. Suppose there are five plants involved, with considerable variation in size, and you are interested in comparing their accident record. To look simply at the total number of accidents in each plant would tell you very little because you don't know how much of the variation is due to differences in size of the work force. The ratio of accidents to working hours, or some equivalent of exposure, eliminates this difficulty and is fully comparable from one plant to another.

A number of ratios are in common use in the business world, such as:

Current ratio: ratio of current assets to current liabilities. Measures the ability of a company to meet its current obligations.

Quick ratio: ratio of current assets less inventory to current liabilities. An even more rigorous test of the company's short-term financial strength.

Debt ratio: ratio of long-term debt to stockholders' equity. Size of the debt in relation to the company's net worth is a direct measure of the leverage being used in its asset structure.

Turnover ratio: ratio of annual sales to average inventory throughout the year. The faster the turnover, the smaller the amount of capital that must be tied up in inventory.

P/E ratio: ratio of stock price to earnings per share. One measure of whether a stock price is relatively high or low.

Profit margin: ratio of net income to sales. A key ratio in the performance of any company.

Many more items could be added to this list, but it is sufficient to indicate that ratios are put to extensive use in the business world.

PERCENTAGES

In preparing material for their own analysis or for presentation to others, executives will often find that nothing is more useful than a simple conversion of data into percentages. Because of the standard use of the decimal system, it is very easy to visualize any number as part of 100.

For example, assume this breakdown of expense in a given plant, with totals expressed in thousands of dollars:

Item	Expense	Percent
Adm. and Sales....	$ 2,114	16.3
Production Cost....	3,412	26.3
Raw materials.......	4,678	36.0
Utilities and Energy	1,768	13.6
Depreciation	1,024	7.9
Total	12,996	100.0

And it is obvious that addition of the percentage column has made it much easier for the reader to gauge the importance of each item, not only in relation to the total, but also in relation to each of the other items.

By the same token, if you were comparing two or more plants in terms of relative expense, it would be useful to convert the data into percentages as in the example above. Again, because percentages are abstract numbers, they are fully comparable for this purpose even though the plants differ materially in size.

Percentages are also very useful in measuring change from one time period to another. For example, assume the following sales in thousands of dollars for five divisions of a company:

Division	Year 1	Year 2	Pct. change
A	2,232	2,147	− 3.8
B	4,113	4,207	2.3
C	3,881	3,957	2.0
D	1,773	1,907	7.6
E	617	743	20.4
Total	12,616	12,961	2.7

Clearly, the percentage column throws the relative change of each division into sharp focus almost at a glance. And it is equally clear that it is not easy to do this by simple inspection of the original data.

Where data are cross tabulated, it is possible to run percentages in three different ways, as illustrated in Table 7.1. These data are from a market survey reporting on women consumers of two competitive products with the women classified by two age groups. The original data are shown in Section I of the table, then percentages are run

TABLE 7.1
Women consumers of two competitive products

	Product	Under 35	35 & over	Total
I	A	242	156	398
	B	197	239	436
	Total	439	395	834
II	A	55.1	39.4	47.7
	B	44.9	60.6	52.3
	Total	100.0%	100.0%	100.0%
III	A	60.8	39.2	100.0%
	B	45.2	54.8	100.0%
	Total	52.6	47.4	100.0%
IV	A	29.0	18.7	47.7
	B	23.6	28.7	52.3
	Total	52.6	47.4	100.0%

by column, then by row, and finally as a ratio to the total. There is really no right or wrong choice here. In each case, the percentages provide helpful perspective.

The column percentages, for example, show quite clearly the preference of the younger group for Product A and the preference of the older group for Product B. The row percentages reveal the heavy dependence of Product A on the younger women and the more even balance of Product B (although the older women are in the clear majority). And finally, percentages of the total show the importance of each subgroup in this particular market segment.

Thus, since all these percentages serve a useful purpose, there is nothing wrong with showing them all. The decision on this, of course, depends on other data that might be involved and the amount of detail that is appropriate for the report as a whole.

You may wonder if there is any rule on how many decimal points to show in a percentage. There is no such ironclad rule but, in general, I would recommend one decimal point. An exception to this would be with certain survey or sampling data which are simply not that accurate and thus should be rounded to whole numbers to avoid misleading the reader.

SUMMARY

Ratios are widely used in the business world. In a very real sense, the ratio of one number to another is often more informative than either of the original numbers taken alone. And the ratio, being an abstract number, is comparable to others of like nature, irrespective of any size differences of the original numbers involved. This makes it possible to apply and draw inferences from ratios in a wide range of circumstances.

Ratios in percentage form are a commonplace in the business world, as elsewhere. Because parts of 100 can be readily visualized, percentages can throw relationships into instant and sharp focus. For this reason, they should always be considered in the analysis of any data and its presentation to an executive audience.

CHAPTER 8
Compounding

The principle of compounding enters into the solution of many problems in business, and it may be worth taking a small refresher course on the fairly easy arithmetic involved.

Start with the value of one dollar compounded annually at some rate of interest. The formula for this is:

$$\$1 \text{ Value} = (1 + \text{Rate})^N$$

where N is the number of years and Rate is in decimal form; i.e., 10 percent = .10.

Thus, the value of one dollar compounded at 10 percent annually for ten years is 1.10 to the 10th power, which is $2.59. To solve this with a hand calculator, simply enter 1.10, then press the power key, y^x, then enter 10 and touch the (=) key.

Incidentally, if you don't already have one, you should buy one of these little calculators, preferably a scientific model that will do power, root, and log computations. Ones are available that cost less than $20, will fit in your pocket, operate on solar power and never need recharging, and yet have awesome computing capability. Remarkable technology in our world today!

The compounding formula can produce some results that are a bit surprising. One example is found in the continuous war between banks and savings institutions in which they promise various things in the interest they will pay for deposits.

On something like a one-year certificate of deposit, for example, the original practice was to calculate and pay the interest at the end of the year. But then some enterprising bank decided to advertise that it would pay interest compounded every month. Then others decided to go that one better and said they would compute interest every day.

To analyze this, the formula needs a slight modification:

$$\$1 \text{ Value} = (1 + \text{Rate}/F)^F$$

Here F is the frequency of compounding throughout the year and Rate is the same annual rate as before, in decimal format.

Now, to see what all this means to the saver, assume the investment of $1,000 for one year. Here is the interest earned at a 10 percent annual rate, compounded at various frequencies throughout the year:

Frequency	Interest
Annual	$100.00
Monthly	104.71
Daily	105.16
Hourly	105.17
Infinite	105.17

So, going from annual to monthly compounding adds $4.71 to the interest accumulation. Going from monthly to daily adds only $0.45, and from daily to hourly only one penny!

And the hourly compounding is the best you can do, because any further frequency—even every billionth of a second— adds no more to your interest total!

It is very easy, incidentally, to compute the maximum effect of such compounding. To do this, simply take the antilog of the rate in decimal format. On the calculator enter .10, then (inv) and (ln), and the answer appears: 1.1051709 (which is the value of one dollar plus accumulated interest after one year of compounding).

When amounts grow with compounding at a constant rate, the increments grow larger and larger, and finally can be very large indeed. If, for example, one of your ancestors 300 years ago had been sufficiently astute to invest $1.00 in a Swiss bank to earn interest at 10 percent per annum, this estate today (assuming no taxes on the Swiss bank account) would be worth:

$$2,617,000,000,000$$

a tidy sum even in comparison with our national debt!

This also explains how insurance companies today (many of which now seem to be investment houses, with insurance as a side fringe benefit) can offer what appear to be such fabulous rates of return.

Zero coupon bonds also illustrate the point. A thirty-year bond at 10 percent interest accumulates a $17.45 final payment per dollar invested at the beginning. Not a bad investment for young people to consider!

COMPOUNDING IN REVERSE

Sometimes you need to reverse the compound interest formula to answer certain questions. For example, assume that your division increased its sales from $12.3 million to $17.9 million in the last five years. What has been the annual compound growth rate? The answer is found in this formula:

$$\text{Growth Rate} = (\text{New}/\text{Old})^{(1/N)} - 1$$

When you divide 17.9 by 12.3 you get 1.4553, and the fifth root of this (or this number to the 0.2 power, which is the same thing) is 1.078. Subtracting one you get .078, or 7.8 percent, as the annual compound growth rate.

This can then be compared directly with the growth rate of the company as a whole, or the industry, or any other standard that may be relevant. Once again, all this procedure requires is two or three entries in a little hand calculator, and it is a very useful routine to know. I must warn you, however, that there are some dangers of interpretation in this shortcut procedure, which are explained in the chapter on Compound Growth Rates.

FRAUD IN COMPOUNDING

When numbers grow at a constant rate they form what is known as an *exponential* series, and such a series can grow rather quickly into astronomical size. This can be quite deceptive and forms the basis for a number of fraudulent schemes. The chain letter is a good illustration. It seems very easy to send a dollar or so to prior names, strike one, add your own, and then send the letter on to five others to repeat the process.

But the number of letters that must be mailed in each consecutive wave to keep this process going, if N represents the wave of mailing, is five to the Nth power. Thus, on the 14th such wave, letters would have to be mailed to more than 6 billion recipients, which, of course, is well in excess of the total world population.

The same principle is involved in so-called *Ponzi* schemes of one kind or another in which investors are promised fabulous rates of return on their investment, such as doubling every three months. They are so enthralled by this that they often continue to add to their initial investment until the day, alas, when their golden dream is shattered and the promoter has disappeared with all their money!

Notwithstanding the obvious fallacy in such schemes, they continue to crop up in a variety of forms to the sorrow of their victims, who are often quite intelligent and should know better. Indeed, I know of one rather famous case where some otherwise astute bankers not only recommended the scheme to their customers, but happily invested their own money as well. The day of reckoning was a very sad one indeed for all concerned.

FORECASTING DANGERS

In a more normal context in the business world, it is not uncommon to see revenue and profit projections based on some compound growth rate, perhaps extrapolated from prior years. As a general rule, all such projections should be viewed with great suspicion.

Nothing in this world continues to grow at a constant compound growth rate for very long, and projections of this kind can be highly misleading. But this is not always understood, and forecasts of this kind seem quite objective and rational and can easily be accepted as such. I have seen it happen with disastrous results for those who built their entire business plan on such a projection.

SUMMARY

The law of compounding turns up in a great variety of ways in the business world, sometimes for good and sometimes for bad. Every executive should understand the simple arithmetic involved and know how to use, and not be misused by, this rather fascinating numerical concept.

CHAPTER 9
Present Value Analysis

In a great variety of business problems, you must deal with the flow of money over time, and a special technique is required for this type of analysis. Clearly, since the value of a future dollar is not the same as that of a dollar now in hand, a simple summation of funds to be provided at different time periods is a classic mixture of apples and oranges.

What is needed is some means of valuing all these future money flows in terms of some common unit, and the logical unit to use for this purpose is the value of the dollar at the time of the analysis, which is to say its present value.

Once this is settled, the only other thing required is to specify some interest rate to be used in equating all these values. Here is a very simple example. Suppose you have a note coming due in five years that will pay you $10,000. What is its present value to you?

Assume further that you can invest in a certificate of deposit that will pay 10 percent compounded annually with a five-year maturity. What would you pay for a $10,000 certificate of this kind?

There is a single answer to both questions. The certificate would cost you $6,209, which at 10 percent per annum will grow to $10,000 in five years, and this accordingly is the present value of the note at this rate of interest.

In other words, it would make no difference to you if the five-year $10,000 note were exchanged for a five-year certificate of deposit with the same maturity value, which could be purchased today for $6,209. As far as you are concerned, the values are identical, and thus the $10,000 note is really worth only $6,209 in terms of current, or present, value. This is the simple concept involved in present value analysis, and the arithmetic is equally simple.

In the example cited, first ask the question of how one dollar will grow in five years at 10 percent per annum. The answer to this is 1.10^5, which is equal to 1.61051. When this is divided into 10,000, you get $6,209 as the base, or initial, amount.

Another way of looking at this, which may even be easier to grasp, is that the value of $1.00, at this rate of interest over five years, deteriorates to $0.6209 in terms of present value; and when you multiply .6209 times the $10,000 future value of the note, you get $6,209 as the present value.

SERIES OF PAYMENTS

This example involves only one payment in the future, but exactly the same procedure is followed to find the present value of the entire stream when there is a series of such payments. In actual practice, this is where present value analysis can be most useful. Here is an example to show how this works.

Assume that your company is disposing of a surplus plant and must choose between two offers. The first is a cash payment of eight million dollars. The other is a payment of two million dollars at the end of each year for the next five years. Which offer should be chosen?

Clearly, in order to compare the cash offer with the installment alternative, the latter must be converted to its present value equivalent. The key question here is the interest rate to use in the analysis.

If you select 8 percent as the interest rate, the installment payments have a present value of $7,985,000, or almost exactly the same as the $8 million cash offer, as demonstrated in Table 9.1. So it is a matter of relative indifference, in terms of present value, which one is chosen over the other. In actual practice, of course, the cash offer would normally be preferred in this situation if there is even the slightest element of risk in the installment payout, which usually is the case.

In any event, the main thing to learn from this present value analysis is that the extra $2 million in the installment offer is essentially fictitious in terms of present value and should be disregarded accordingly.

The format of Table 9.1 is very valuable in presenting the results of a present value analysis. Some years ago I programmed my own computer to use this format in solving problems of this kind because it is an excellent way to present the material to others.

First it shows the discounted value of each dollar in the future, based on compounding at the specified rate of interest. Then it shows

TABLE 9.1
Interest rate (%) 8

Year	Value of $1	Annual return	Present value	Cumulative
1	0.9259	2,000	1,852	1,852
2	0.8573	2,000	1,715	3,567
3	0.7938	2,000	1,588	5,154
4	0.7350	2,000	1,470	6,624
5	0.6806	2,000	1,361	7,985

the result of multiplying this by each future payment to arrive at the discounted value of each such payment, and then the sum of these to arrive at the present value of the entire stream of payments.

Often many executives are given a present value figure that was developed by someone else and then they make decisions based on that figure with only a vague idea of what it really means. They simply accept the figure as a matter of faith. That does not seem to be the way to make multimillion-dollar decisions. This is why I like the format of the table, which exposes the entire analytical procedure and makes the concept clear to all users of the material.

For example, anyone looking at the table might well wonder what the present value of the installment payments would be at other rates of interest. This is worked out in Table 9.2 for two other rates: 6 percent and 12 percent respectively with significant changes in the net result.

With interest at 6 percent, the present value is $8,425,000, versus $7,210,000 at 12 percent. In the first case, the installment payments have a present value well above the $8 million cash offer; in the second case, well below. Here different assumptions about the rate of interest lead to two very different decisions.

It is evident, therefore, that the choice of interest rate is critical in any present value analysis, with widely varying results associated with changes in this assumption. This is all the more reason, of course, why decision makers should understand the analytical procedure involved and be duly cautious about accepting a single-figure result provided by others.

Please note, also, that there is no one magical interest rate number that is appropriate for all situations at any given time. What may be

TABLE 9.2

Interest rate (%) 6

Year	Value of $1	Annual return	Present value	Cumulative
1	0.9434	2,000	1,887	1,887
2	0.8900	2,000	1,780	3,667
3	0.8396	2,000	1,679	5,346
4	0.7921	2,000	1,584	6,930
5	0.7473	2,000	1,495	8,425

Interest rate (%) 12

Year	Value of $1	Annual return	Present value	Cumulative
1	0.8929	2,000	1,786	1,786
2	0.7972	2,000	1,594	3,380
3	0.7118	2,000	1,424	4,804
4	0.6355	2,000	1,271	6,075
5	0.5674	2,000	1,135	7,210

correct for one company may not be proper for another. To assume a relatively high interest rate could well be appropriate for a company with a strong need for immediate cash, but quite inappropriate for another company with a large cash surplus.

If a company is heavily in debt and a poor credit risk, for example, the interest rate it must pay on borrowed money for operating needs can be quite high. In another company with excess cash, the value of money is simply what it can earn on short-term investments, normally at a relatively low rate.

Thus, the interest rate selected in a present value analysis is a matter of executive judgment in each case. It should reflect a careful evaluation of the relative need for cash, now and in the future, as well as the overall level of interest rates. This is fundamental, and you should always remember that the usefulness of any present value analysis is wholly contingent on the quality of this judgment.

One further word of caution. Tax consequences have been ignored in this discussion, but in real life they must be considered, and there may be a significant difference in this respect between installment payments and a current cash payment. In most instances, present value analysis should be based on after-tax payments; and it should always be clearly specified if pre-tax payments are used instead.

SUMMARY

Present value analysis is a simple but ingenious way to equate future and present dollars. Without such a device, there would be no proper way to evaluate a stream of future payments.

Although this is a standard procedure in the business world, a surprising number of executives are less than knowledgeable about it and are handicapped accordingly in making proper use of the results. The interest rate assumption is critical and always requires careful executive judgment.

Apart from its own value as an analytical device, the present value concept is a key element in arriving at something called the *internal rate of return:* perhaps the most powerful single figure in the business world and the subject of the next chapter.

CHAPTER 10
Internal Rate of Return

The higher you rise in executive rank, the more often you are likely to encounter something called *internal rate of return*.

This may well be the most powerful single statistic in the corporate world because it underlies—and provides the basic rationale for—the expenditure of billions of dollars in capital outlays, acquisitions, and other investments.

Yet I would venture the guess that very few of the executives, who must base decisions on a single number of this kind, could tell you precisely how it is calculated.

They understand roughly what it means—the name itself is at least partly self-explanatory—but this knowledge is far from precise. And this is a handicap because, with a greater understanding, they would be able to ask questions that should be considered, instead of simply accepting the number on faith.

Typically, of course, those recommending a substantial capital investment of some kind are the ones who calculate the projected rate of return and submit it along with their request for the given appropriation.

It seems most imprudent for the actual decision maker to accept such a number without knowing exactly how it was calculated. This is not to imply any intent to deceive, but simply that those who did the actual work may not have used the right assumptions or the proper analysis. Consider an actual example. You are thinking about buying a machine to improve productivity. It costs $10,000 and has a useful life of five years with no salvage value. Here are your expected savings, year by year:

Year	Savings
1	2,000
2	3,000
3	4,000
4	5,000
5	6,000
Total	20,000

68 BASIC BUSINESS STATISTICS FOR MANAGERS

Now, is this a good investment or not, compared with other ways in which you might invest this money? Since you invest $10,000 and get $20,000 back, you have a profit in total dollars—but the key question is: how good is the profit in the form of expected return on the total investment?

How do you answer this? Well, if you remember the explanation of *present value*, you will recall a basic lesson that you learned: the value of money varies according to the time period involved related to some interest rate assumption.

You saw how you could take any stream of payments over time, apply a given interest rate, and compute what all those payments are worth in terms of current, or present, value.

Now, all you need do is turn that procedure around. You know what the present value is, namely the amount of the investment and you know the stream of payments. So all that remains is to find the interest rate that will convert the stream of payments into a present value that equals the amount being invested.

And this specific rate is the internal rate of return for this particular investment.

In this example, the interest rate turns out to be 23.2919 percent, as you can see in the table below:

Year	Savings	$1.00 Compounded	P. value $1.00	Present value
1	2,000	1.233	.811	1,622
2	3,000	1.520	.658	1,974
3	4,000	1.874	.534	2,134
4	5,000	2.311	.432	2,164
5	6,000	2.849	.351	2,106
Total	20,000	-	-	10,000

The calculations should be familiar by now. The third column shows the value of $1.00 compounded at the given interest rate, and the fourth column shows the present value of each future dollar based on this interest rate (which, you will recall, is the reciprocal of the compounded value).

Present value of the savings each year is derived either by dividing actual savings by the compounded value per dollar or by multiplying by the present value of each future dollar. These values then add up to the amount of investment.

So the concept of the *internal rate of return* is no more complicated than this. It is simply the interest rate that you are earning on your investment when you take into account the stream of payments or benefits that you expect to receive.

Put another way: the analysis tells you that your investment will be recovered from the stream of future benefits, even though future dollars are discounted at the specified rate of interest.

HOW TO CALCULATE

If it is all this simple, wherein does the complexity lie? You may have noted by this time that I have omitted one thing. You know what the interest rate—and thus rate of return—was in this example, but you do not know where the figure came from.

And the reason is that there is no simple equation that will give you this answer. This is one of the problems in mathematics that can only be solved by successive approximation.

You must first make a guess at the interest rate and calculate the present value on that assumption. Seeing that result, you make a better guess and calculate again. And so on and on, until you get close enough to the present value (amount of investment) to be satisfied with the result.

This can be a very tedious and time-consuming task, and so it was not too many years ago when calculators ground away at these successive approximations.

But modern technology has really put an end to all this. With lightning speed, calculators and computers can run these onerous calculations and quickly give you the answer accurate to several decimal points. The task, indeed, is reduced to no more than listing the numbers, while the machine does all the rest of the work.

So the only complexity in the *internal rate of return*, the method of computation, has been eliminated by modern machines, and no one needs to be daunted by that any more. The key fact is that the concept itself is simple, straightforward, and easy to understand.

BASIS FOR CALCULATION

In actual practice, the decision maker should pay special attention to the projected stream of savings or payments in any calculation of this kind. In a large company with these data coming from many sources, it pays to be quite skeptical on this score.

On this I can speak with the considerable experience of twenty years as chief executive officer with the responsibility of approving these requests on a daily basis, in a total amount over this period of perhaps half-billion dollars or so.

In most instances, the stream of benefits should be on the basis of cash to be returned to the company on an after-tax basis, but I have found many variations from this in actual division reports. Quite often, the benefit stream used in the calculation takes no account of depreciation, taxes, or tax credits—and this, of course, can produce a result that is vastly different from true cash return.

Beyond this, of course, is the question of reliability. How much faith should be attached to the schedule of expected benefits. Again, prudence suggests that those recommending capital outlays may be a bit optimistic in estimating the anticipated return.

This natural prejudice is strengthened by the knowledge that no one, after the fact, is ever likely to check on the accuracy of the projection. In theory this should be done, but most companies are much too busy with current problems to indulge in this kind of historical research, even if corporate records could give them the answer.

This is all the more reason why all the assumptions involved in the analysis of this kind should be reviewed with considerable care in the final decision-making process.

SUMMARY

The apparent simplicity of any *rate of return* figure can be quite misleading unless one has explicit knowledge of the elements that enter into the calculation.

But the concept itself is both simple and elegant. How else could you express the return on any investment, in the form of any stream of future benefits, in one summary number for easy comparison with the return on other investments or with some minimum standard of return for all investments in a company?

In any event, if you have learned the lesson of this chapter, you are miles ahead of most executives on this subject. And you may find the lesson especially helpful if you are involved in preparing requests for capital outlays, and even more so if you are in the position of recommending or approving such requests.

CHAPTER 11
Probability

The laws of probability are involved in many intricate ways in the world around us and touch our lives every day. The mathematical rules that apply to probability are quite complex, and entire books have been written on the subject.

This brief chapter merely provides an overall glimpse of what probability is all about and tries to give you some grasp of basic principles. This degree of knowledge is really quite vital for anyone who wants to understand how numbers operate in the real world.

This is especially true for executives because risk of one kind or another is involved in almost every business decision, and there is a constant need to weigh the relative probabilities involved in the decision-making process.

The probability of a certain outcome of a given event can be defined as the relative frequency of that specific outcome among all events of the same kind. Thus, the probability that a certain outcome will occur is some fraction between zero and one. A value of zero means that it will never occur; a value of one that it is certain to occur. A specific probability is some value in between.

For example, take the first ten letters of the alphabet. What is the probability of selecting at random (each with an equal chance of being selected) any one of these letters? The answer, clearly, is one in ten, or 1/10. Similarly, the probability of selecting at random the ace of spades from a deck of cards is 1/52. And so on.

Once again, select one letter at random from the first ten letters of the alphabet. What is the probability that you will draw either B or D? Here the probability of selecting either letter is 1/10, and the probability that at least one of the two will be chosen is 1/10 + 1/10 = 2/10. This illustrates the addition law of probability:

> The probability of getting at least one of two or more mutually exclusive occurrences is the sum of their individual probabilities.

Mutually exclusive occurrences mean that both cannot happen. If you select a marble from a jar with only red and white marbles, the one selected must be either red or white: it cannot be both.

Now, look at another problem. In a first draw from the ten letters, you select one. After it is replaced, you select another letter in a second draw. What is the probability that a specific letter, such as C, will be selected in both draws?

In each draw, the probability of selecting C is 1/10, and the probability of selecting C in both draws is 1/10 x 1/10, or 1/100. This is the multiplication law of probability:

> The probability that two or more independent events will have certain outcomes is the product of the probabilities of these outcomes.

Independence, in this context, means that the probabilities of various outcomes of each event are the same, whatever the outcome of the other events.

It may be helpful to look at this in a diagram, and for this purpose consider the probability of selecting C from among the first five letters of the alphabet in each of two draws:

Trial II	Trial I				
	A	B	C	D	E
A	0	0	0	0	0
B	0	0	0	0	0
C	0	0	X	0	0
D	0	0	0	0	0
E	0	0	0	0	0

Here are all the possible combinations of two letters in the two draws, 25 in all, and since only one combination meets the success criterion of selecting C in both draws, the probability of this joint occurrence is 1/25.

Since the probability of selecting C is 1/5 each time, the multiplication law specifies that the probability of selecting C on both draws is 1/5 x 1/5, or 1/25, as the diagram confirms.

The diagram illustrates the general rule in probability analysis. If all outcomes are equally likely, simply count those that represent success as a ratio of all possible outcomes to find the overall probability of success. This principle not only has the virtue of conceptual clarity, but can also lead to the solution of some rather complex problems.

PROBABILITY OF SUCCESS

For example, assume that you are considering a business project with two key elements essential to its success. Further, assume that key managers have agreed upon an 80 percent chance of success in each element. What is the probability that the project will succeed?

Once again a simple diagram can help find the solution to this problem. It is comparable to the five-letter example involving two draws. One letter must represent failure in each case to equal the specified 80 percent probability of success in each element. Assume that letter to be C in each draw. Then you have:

Trial II	\	Trial I			
	A	B	C	D	E
A	0	0	X	0	0
B	0	0	X	0	0
C	X	X	X	X	X
D	0	0	X	0	0
E	0	0	X	0	0

Here again are all possible combinations of the five letters in the first trial and the second trial. Whenever a combination contains one or more C's it is a failure and is marked in the diagram with an X. There are 9 such combinations that represent a failure versus 16 that represent a success out of a total of 25 possible events.

So the probability that the project will succeed is 16/25 or .64. You could have obtained this answer directly, of course, by using the multiplication rule, with the product of the individual probabilities being 4/5 x 4/5 = 16/25, or .8 x .8 = .64.

There is no objection, incidentally, to expressing probabilities in percentage form—64 percent in this case—and indeed this may often be the preferred format.

In terms of real life, we can expand on the preceding example to see some of the difficulties in carrying out a complicated mission such as a space flight. Assume there are 1,000 key parts in the ship, with the probability of only 1/1000 that any one of these will fail. What is then the probability that all will function correctly?

The answer, found by the multiplication rule, is .999 multiplied by itself 1,000 times, which is equal to .3677. In other words, even with this high degree of precision in each part, the odds are only about one in three that all will function correctly in a given flight. It is not surprising, therefore, when failures of one kind or another do in fact occur.

ERROR FREQUENCY

The same principle explains why executives can expect increasing problems as their supervisory span grows to a degree that seems quite disproportionate.

For example, suppose that you are responsible for a single division and that the odds are about even that you will get by from week to week without a serious mistake or problem, which is pretty good performance in the real world.

Then you are promoted to look after ten divisions. Each is operated just as efficiently as the first. Now, what are the odds that you will have no serious mistakes or problems in a given week? The answer: about one chance in a thousand, which is to say virtually no chance at all.

The multiplication rule provides this answer. Simply take for each division the probability of success in a given week (defined as no serious problems or mistakes) and multiply it by itself ten times. This is .5 to the tenth power, which is equal to .0009766 or less than 1/1000.

You may not be pleased with this finding, but it is well to keep it in mind as you move up the executive ladder. The higher you climb, the greater your need for fortitude to bear the slings and arrows of outrageous probability.

In brief, as the scope of your authority increases, you not only will take on greater responsibility but, almost certainly, will find that you have a much larger and more persistent pattern of trouble to contend with.

Having gone the entire route from managing a relatively small company that grew into a fairly large company, I can testify that this law works in exactly the same way as the example indicates.

Many years ago, my close associates fell into the habit of dropping by my office on Friday afternoon to brief me on any significant events of the week that had not already come to my attention. As our company grew larger, it was a rare Friday indeed without its full quota of bad news to unbrighten the weekend.

The bright side, of course, is that personnel consultants do take all this into account in their point rating systems that establish salary guidelines, and your compensation should grow accordingly as you take on additional responsibility and the associated law of compound probability!

Incidentally, all this suggests a fallacy in the way many people view probability. After a considerable span of bad luck, they reason that the odds heavily favor a switch to good luck. But, unhappily, this is not so.

For example, if the odds of getting heads on the toss of a coin are 1/2, the probability of getting a head on ten consecutive tosses is quite small: 1/2 to the tenth power, or 1/1024. But this does not mean that if you have tossed nine heads in a row, the odds are 1,024 to one against a head on the tenth toss. That probability, on the tenth toss or any other, remains precisely 1/2. Sad, but true!

None of this, of course, contradicts the *Law of Large Numbers*, which specifies that sample proportions will diverge less and less from the population average as the size of the sample increases. If the true probability is 1/2, the more coins you toss, the closer you are likely to get to that result as an overall average.

COUNTING VARIOUS OUTCOMES

So much for this digression, and back to the system of counting various outcomes to determine probability. There is an excellent illustration in two dice and the game of craps. Assume, of course, honest dice with each side having an equal chance of turning up on each roll.

The question is: how many ways can the two dice be combined to come up with a given sum of the two? Here are the possible combinations:

76 BASIC BUSINESS STATISTICS FOR MANAGERS

Sum	Combinations						Pairs
2	1-1						1
3	1-2	2-1					2
4	1-3	2-2	3-1				3
5	1-4	2-3	3-2	4-1			4
6	1-5	2-4	3-3	4-2	5-1		5
7	1-6	2-5	3-4	4-3	5-2	6-1	6
8	2-6	3-5	4-4	5-3	6-2		5
9	3-6	4-5	5-4	6-3			4
10	4-6	5-5	6-4				3
11	5-6	6-5					2
12	6-6						1
Total							36

In the first role of dice, you lose if you throw a two, three, or twelve, and you win if you throw a seven or eleven. By adding up the number of pairs related to these outcomes, you can see that you have 4/36 chances to lose and 8/36 chances to win on this first roll.

Say that you throw a five instead. Now, if you throw another five before you do a seven, you win; otherwise, you lose. Your chances of getting a five are 4/36 versus 6/36 for the seven, so the odds are four to six that you will win or six to four that you will lose.

And so it goes. Again, if the dice are honest, the table gives you the exact odds of getting any specified sum of the two dice versus any other sum, and you can see that the system of counting all possible outcomes is a very effective device in the analysis of this and similar situations.

UNUSUAL RESULTS

The laws of probability can illuminate and explain many strange phenomena.

A good example is the experience that most of us have had while dining out, where musicians, having serenaded the birthday of one guest, repeat the performance at another table, and everyone remarks how weird it is that two people in the room have the same birthday.

Yet the fact is that this occurrence is not unusual, but, on the contrary, to be expected. Here is why:

Start with the first person in the room. The chance that he or she will have one of the 365 days as a birthday is a certainty, or 365/365. The chance that the second person will have a different birthday is 364/365, because there are only 364 days left. The chance that the third person will have a different birthday from the other two is 363/365, and so on.

Thus, to get the probability of a different birthday for all members of a group, multiply 365/365 by 364/365 for two people, then multiply that product by 363/365 for three people, and so on. When there are 23 people, the probability of that many individuals having a different birthday is .493, which means that the probability is .507 (or the odds are just about even) that at least two will have the same birthday. If there are 50 people in the room, the odds in favor of two having the same birthday are about 32 to 1; and for a group of 100 people, the odds are about 3,300 to 1!

Usually, no one will believe you if you tell them this because it violates all their rules of common sense, and thus it may be better to give a little demonstration in the form of a friendly wager, at highly favorable odds, at some point before the serenading begins.

SUMMARY

Most business decisions require some assessment of probabilities, and executives should understand the simple but basic rules explained in this chapter.

The laws of probability have a logic of their own and often lead to strange conclusions that most of us could not possibly deduce by so-called common sense reasoning alone. Failure to understand this can lead to some very serious errors indeed.

CHAPTER 12
Permutations and Combinations

At some time in your past you undoubtedly learned something about the law of permutations and combinations. At the time, it probably seemed no more than just another mathematical abstraction having nothing to do with any real life situation.

This is not correct. This little law on how things can be arranged and combined is, in fact, a tremendous help in solving many problems, and business executives should know how to use it accordingly. This is especially true when probabilities are involved and the methods outlined in the preceding chapter are needed.

When the numbers are small, it is easy enough to count all the possible outcomes as well as those considered to be favorable in order to calculate the probability of the latter. But this becomes quite impractical with larger numbers, and we need a more efficient method.

For example, consider the probability of getting thirteen spades in a bridge hand. For this, you need to know the total number of possible hands, and to count them individually would be quite a task. Indeed, even if you could count them at the rate of one per second, it would take 201,350 years to complete the job!

There is a much easier way to get at this, but first you must understand factorials and scientific notation.

The factorial of a whole number is the product of that number and all preceding numbers. To indicate a factorial an exclamation mark is listed after the number, so that 5 factorial is represented by 5! And the rule just given means that $5! = 5 \times 4 \times 3 \times 2 \times 1 = 120$. It should be noted, incidentally, the 0! is always equal to 1.

This is simple enough, but factorials grow quickly into some very large numbers indeed. For example, factorial 10 is some 3.6 million, and factorial 20 is greater than two followed by 18 zeros. When we deal with numbers of this size, we need some way to express them in a more compact form.

Permutations and Combinations 79

This is accomplished by way of *scientific notation*, in which the number is shown with an exponent that represents some power of 10 or 1/10. For example, 125 can be written as 1.25E+2, which simply means 1.25 multiplied by 10^2 or 100. Similarly, .05 can be written as 5E-2, which means five multiplied by 1/10 to the second power, or 1/100.

One simple way of remembering this is that if the number following E is positive, the decimal point should be moved this many places to the right; if the number is negative, the decimal point should be moved this many places to the left.

Thus, when you see a number like 3.5E+6, you can see instantly that it represents 3.5 million. Change the E number to 9 and it then represents 3.5 billion. And so on.

All this is very handy, especially since computers are smart enough to deal with numbers in scientific notation just as readily as they can with all other varieties.

PERMUTATIONS

Now you are ready to consider the subject of this chapter. Start with permutations, which is the number of possible ways to arrange n objects taken r at a time, written as $_nP_r$. The formula for this is:

$$_nP_r = \frac{n!}{(n-r)!}$$

To illustrate this, take the simple case of the first three letters of the alphabet taken two at a time. The possible arrangements are: AB, BA, AC, CA, BC, and CB, which corresponds with the formula result:

$$_3P_2 = \frac{3 \times 2 \times 1}{1} = 6$$

Similarly, five letters taken three at a time can be arranged in 60 different ways:

$$_5P_3 = \frac{5 \times 4 \times 3 \times 2 \times 1}{2 \times 1} = \frac{120}{2} = 60$$

80 BASIC BUSINESS STATISTICS FOR MANAGERS

In each case, note that the (n-r)! divisor cancels out that many numbers to the right in the numerator, with the result that the last calculation, for example, can be abbreviated to 5 x 4 x 3.

The Word Jumble wherein two arrangements of five letters and two of six letters must be rearranged to come up with four target words, is a popular feature in newspapers.

The formula above reveals that with five letters taken five at a time, there are 5 x 4 x 3 x 2 x 1 possible ways to arrange the letters, or 120 in all. Make it six letters taken six at a time, and the number of arrangements jumps to 720—a much more difficult Word Jumble to solve.

By the same token, a truck driver with deliveries to make to six customers has no less than 720 different routes to choose from. While most of these would clearly be inefficient, this does illustrate that for a fleet of trucks serving several hundred customers the best routing pattern for the entire fleet is no easy problem to solve.

The simple formula for permutations can provide a very quick and sharp perspective on the dimensions involved in a variety of situations.

NUMBER OF COMBINATIONS

The law of permutations has to do with the number of ways various elements may be arranged, which is very different from how those elements may be combined, where the order of arrangement is of no significance.

The formula for the number of combinations of n things taken r at a time is:

$$_nC_r = \frac{n!}{r!(n-r)!}$$

This is same as the permutations formula, except that r! has been added as another divisor. Thus, the number of combinations of five letters taken three at a time is:

$$_5C_3 = \frac{5!}{3!\,2!} = \frac{120}{12} = 10$$

This is the same as the number of permutations (60) divided by 3! Computers can run the permutations and combinations formulas—even with very large numbers involved—in a flash. In the example of the total number of possible bridge hands, there are 52 cards selected 13 at a time, and the number of different combinations is:

$$_{52}C_{13} = \frac{52!}{13!(52-13)!} = 6.35E + 11$$

This translates to 635 billion possible hands. You can see why it would take a long time to count them all even with a very fast counting procedure, and why the probability that one player would receive any one specific hand is infinitesimally small.

All this assumes, of course, that there is an equal chance of selecting each hand, and poor shuffling of cards or the presence of a card shark could change these odds materially. Indeed, this caution is appropriate for every application of the laws of probability to the real world. Where we assume that various outcomes have an equal likelihood of occurrence, we must be quite certain that this assumption holds true. If not, real world results can be quite different from those determined by the theorems of probability.

BUSINESS APPLICATIONS

The law of combinations can often illuminate many problems in the business world. Suppose, for example, that you must choose five members to serve on a task force from a pool of ten executives. This is not as easy as it might appear, because you have 252 separate possibilities to consider. The arithmetic for this, of course, is (10 x 9 x 8 x 7 x 6) /(5 x 4 x 3 x 2 x 1).

Similarly, suppose you have a business plan with ten separate contingencies and you wonder what would happen if two of these were to occur simultaneously. The law of combinations tells you that you have 45 possibilities to consider—(10 x 9) /(2 x 1)—which is a very formidable analysis indeed!

These examples are vivid proof of how essential executive judgment can be in solving many business problems, by setting guidelines and criteria that eliminate many of the alternatives that would otherwise require attention.

Senior executives can become quite talented at this, often seeming to arrive at these answers by sheer intuition, which, of course, can often be the highest form of judgment.

SUMMARY

This has been only a brief survey of this overall subject, but the simple principles that have been discussed are sufficient to give you some grasp of the richness and power of these analytical methods.

In business, as in many other fields of activity, these principles can often clarify thinking and contribute to the solution of many practical problems, and all executives are at some disadvantage if they do not understand at least the fundamentals of the subject.

CHAPTER 13
Statistical Inference

This chapter deals with the magic kingdom of statistical inference where miracles can be achieved in the accumulation of knowledge by translating small pieces of data into findings of broad generality.

Public opinion polls, for example, are commonplace these days and, if you are like most readers, you doubtless wonder how the answers of a thousand or so respondents can be projected to the entire adult population of the United States. And you may wonder even more how it is possible to set precise error limits on such a projection.

Common sense tends to reject this inference of so much from so little, and thus it is not surprising that many should view the entire process with great suspicion. Still others, who reluctantly concede that the procedure seems to work, nevertheless look upon it as some kind of magical trick, cloaked in mystery.

It is the purpose of this chapter to unravel the mystery and lay bare the mechanism that underlies the statistical magic in this procedure. It is important to understand this because many scientific and business decisions must be made on the basis of sampling data of one kind or another.

In studies of market share, consumer preferences, advertising recall, and the like, and in such areas as quality control, executives must deal with inferences based on samples, and they need to know how to gauge their reliability and significance.

EXPERIMENT

First, start with an experiment of tossing two coins and counting the number of heads, where heads and tails are equally probable on any given toss. You can find the expected number of heads, either 0, 1, or 2, by simply listing all the possible combinations of the two coins:

T T
H T
T H
H H

This leads to the following tabulation:

Heads	Frequency
0	1
1	2
2	1
Total	4

This shows that, in consecutive tosses of two coins, the theoretical expectation is zero heads one-fourth of the time, one head two-fourths of the time, and two heads one-fourth of the time.

By the same process, you can find the expected number of heads in a toss of three coins as follows:

Heads	Frequency
0	1
1	3
2	3
3	1
Total	8

You see that, in a toss of three coins, the probability of getting zero heads is 1/8, the probability of one head is 3/8, and so on. And you can work out similar probabilities, using the same process, for whatever number of coins you may wish to toss.

In the case of four coins, for example, the same enumeration process leads to this table:

Heads	Frequency
0	1
1	4
2	6
3	4
4	1
Total	16

Here the probability of getting zero heads is 1/16, the probability of getting one head is 4/16, and so on.

Now, you may be wondering just how accurate all this theory may be in terms of the real world. The frequency distribution shows what to expect, but how much reliance can be attached to this? Is it pure theory, or does it really work out this way in actual practice?

Well, let's put it to the test. Flip four coins and count the number of heads, and then repeat this 1,000 times. A little tedious to do by hand, but my computer can simulate these 1,000 tosses in precisely four seconds! Here are the results:

No. of heads	Theoretical	Actual
0	62	49
1	250	252
2	376	380
3	250	263
4	62	56
Total	1000	1000

and I think you will agree that theory, in this case, is remarkably close to reality!

Please note that you can learn a great deal from this little table of theoretical probabilities. If, for example, you are dealing with a series of four games played by two evenly matched baseball teams, the table shows the odds on any one team winning any specific number of games in the series. Variation is the order of the day, but the theoretical frequency distribution reveals the precise odds involved in each of the possible combinations.

BINOMIAL DISTRIBUTION

In technical terms, what you are dealing with here is a binomial population that is dichotomous: each coin must be either a head or a tail. And the expected number of occurrences in such a population is called a *binomial distribution*.

In the coin example, the probability of a head was 1/2, but this is not true of all dichotomous populations. Assume a jar with 100 red and 900 green marbles thoroughly mixed together. The probability of drawing a red marble from the jar is only 1/10, but this is still a dichotomous population because each marble must be either red or green.

Now, in terms of practical application, consider a poll of 1,000 respondents, selected at random, on which of two candidates they favor. Here again is a dichotomous population: the respondent is asked to choose either candidate A or candidate B.

Assume, in the population being sampled, that there is no difference between the two candidates: that each has an equal chance of being chosen by any respondent. In this case, the most likely poll result would be 500 respondents choosing each candidate. But the coin-tossing examples have shown that many other combinations are possible, and the question is how to determine their probability.

For example, let's say that 520 respondents in the poll chose candidate A. If the expected arithmetic mean is 500, how likely is it that 520 or more respondents would choose the candidate in a sample of this kind?

This is comparable to tossing 1,000 coins, with heads or tails equally likely, and asking for the probability that 520 or more heads will turn up. As in the case of two coins, or three, or four, you could go through the enumeration process for all possible combinations of 1,000 coins and arrive at the theoretical frequencies involved. But this would be a tremendous task and, happily, there is a much simpler short-cut method available to accomplish the same result.

To illustrate this, assume that you work out the theoretical frequencies for the number of heads expected in a toss of 20 coins. These data are plotted in Figure 13.1. Now, expand the number of coins to a larger number, say 100 or so, and the data evolve into the bell-shaped form shown in Figure 13.2: a graph of what is known as the *normal distribution.*

The key point here is that the binomial distribution for a large number of events such as equally balanced coins being tossed can be represented by the normal distribution. And there is a great advantage in this because, unlike the binomial distribution, it is very easy to determine the probabilities involved.

The key characteristic of the normal distribution, depicted by the normal curve, is that it is fully determined by only two statistics: the arithmetic mean of the distribution and its standard deviation. Nothing else is needed to replicate everything shown in Figure 13.2.

In the normal distribution the mean is zero and the standard deviation is one, and comparable values for any sample can easily be translated into their normal equivalent and then be interpreted accordingly.

EXPECTED NUMBER OF HEADS
IN 20 TOSSES OF A COIN

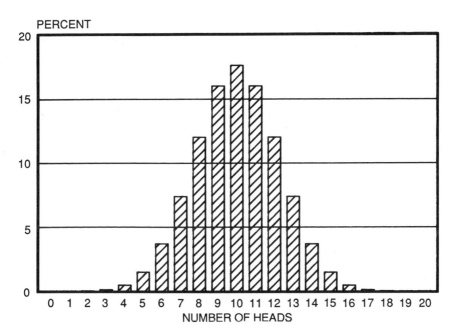

FIGURE 13.1

For example, assume a sample mean of 100 and a standard deviation of 10. Where does the value of 120 fall on the normal curve equivalent? Here 120 is a deviation of +20 from the arithmetic mean, and 20/10 is equal to two standard deviation units. So this specified value is two standard deviations from the mean on the right side of the normal curve.

This is significant because for the normal curve you know the precise area from the mean to any point, measured in standard deviation units, as a percentage of the total area under the curve. And this means that you know the relative probability of any such deviation from the mean.

An abbreviated listing of such values can be found in Table 13.1. Here you see for normal deviates from zero to three (deviation from the mean divided by the standard deviation) the area from the mean to that point, twice that area (which is the area from the mean plus

NORMAL DISTRIBUTION

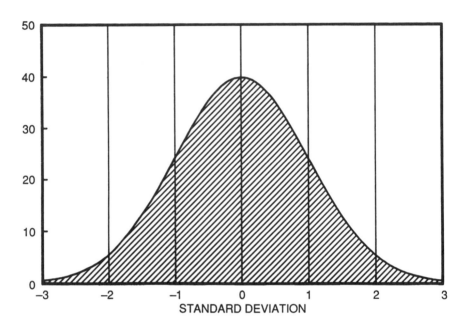

FIGURE 13.2

and minus the normal deviate), and the area that lies outside the latter.

Thus, for a value that lies one standard deviation from its mean, you can see that .3413 of the total area is included. Within an area twice this size, which is that included in the mean plus and minus the standard deviation, .6827 of the total is included. And one minus the latter shows that .3173 of the total lies outside the mean plus and minus one standard deviation. And so on.

Table 13.2 covers a greater range and shows the percentage of the distribution that lies outside the mean +/− each specified normal deviate. Special interest attaches to the 1.96 and 2.58 deviations, where 5 percent and 1 percent of the total distribution, respectively, lie outside these values.

TABLE 13.1
Normal distribution

d/s	Area	2 [Area]	1-2 [Area]
.00	.0000	.0000	1.0000
.10	.0398	.0797	.9203
.20	.0793	.1585	.8415
.30	.1179	.2358	.7642
.40	.1554	.3108	.6892
.50	.1915	.3829	.6171
.60	.2257	.4515	.5485
.70	.2580	.5161	.4839
.80	.2881	.5763	.4237
.90	.3159	.6319	.3681
1.00	.3413	.6827	.3173
1.10	.3643	.7287	.2713
1.20	.3849	.7699	.2301
1.30	.4032	.8064	.1936
1.40	.4192	.8385	.1615
1.50	.4332	.8664	.1336
1.60	.4452	.8904	.1096
1.70	.4554	.9109	.0891
1.80	.4641	.9281	.0719
1.90	.4713	.9426	.0574
2.00	.4772	.9545	.0455
2.10	.4821	.9643	.0357
2.20	.4861	.9722	.0278
2.30	.4893	.9786	.0214
2.40	.4918	.9836	.0164
2.50	.4938	.9876	.0124
2.60	.4953	.9907	.0093
2.70	.4965	.9931	.0069
2.80	.4974	.9949	.0051
2.90	.4981	.9963	.0037
3.00	.4987	.9973	.0027

TABLE 13.2(a)
Normal distribution
Percent of area outside specified deviation

Deviation	Percent	Deviation	Percent	Deviation	Percent
.01	99.2	.41	68.2	.81	41.8
.02	98.4	.42	67.4	.82	41.2
.03	97.6	.43	66.7	.83	40.7
.04	96.8	.44	66.0	.84	40.1
.05	96.0	.45	65.3	.85	39.5
.06	95.2	.46	64.6	.86	39.0
.07	94.4	.47	63.8	.87	38.4
.08	93.6	.48	63.1	.88	37.9
.09	92.8	.49	62.4	.89	37.3
.10	92.0	.50	61.7	.90	36.8
.11	91.2	.51	61.0	.91	36.3
.12	90.4	.52	60.3	.92	35.8
.13	89.7	.53	59.6	.93	35.2
.14	88.9	.54	58.9	.94	34.7
.15	88.1	.55	58.2	.95	34.2

.16	87.3	.56	57.5	.96	33.7	
.17	86.5	.57	56.9	.97	33.2	
.18	85.7	.58	56.2	.98	32.7	
.19	84.9	.59	55.5	.99	32.2	
.20	84.1	.60	54.9	1.00	31.7	
.21	83.4	.61	54.2	1.01	31.2	
.22	82.6	.62	53.5	1.02	30.8	
.23	81.8	.63	52.9	1.03	30.3	
.24	81.0	.64	52.2	1.04	29.8	
.25	80.3	.65	51.6	1.05	29.4	
.26	79.5	.66	50.9	1.06	28.9	
.27	78.7	.67	50.3	1.07	28.5	
.28	77.9	.68	49.7	1.08	28.0	
.29	77.2	.69	49.0	1.09	27.6	
.30	76.4	.70	48.4	1.10	27.1	
.31	75.7	.71	47.8	1.11	26.7	
.32	74.9	.72	47.2	1.12	26.3	
.33	74.1	.73	46.5	1.13	25.8	
.34	73.4	.74	45.9	1.14	25.4	
.35	72.6	.75	45.3	1.15	25.0	
.36	71.9	.76	44.7	1.16	24.6	
.37	71.1	.77	44.1	1.17	24.2	
.38	70.4	.78	43.5	1.18	23.8	
.39	69.7	.79	43.0	1.19	23.4	
.40	68.9	.80	42.4	1.20	23.0	

TABLE 13.2(b)
Normal distribution
Percent of area outside specified deviation

Deviation	Percent	Deviation	Percent	Deviation	Percent
1.21	22.6	1.61	10.7	2.01	4.4
1.22	22.2	1.62	10.5	2.02	4.3
1.23	21.9	1.63	10.3	2.03	4.2
1.24	21.5	1.64	10.1	2.04	4.1
1.25	21.1	1.65	9.9	2.05	4.0
1.26	20.8	1.66	9.7	2.06	3.9
1.27	20.4	1.67	9.5	2.07	3.8
1.28	20.1	1.68	9.3	2.08	3.8
1.29	19.7	1.69	9.1	2.09	3.7
1.30	19.4	1.70	8.9	2.10	3.6
1.31	19.0	1.71	8.7	2.11	3.5
1.32	18.7	1.72	8.5	2.12	3.4
1.33	18.4	1.73	8.4	2.13	3.3
1.34	18.0	1.74	8.2	2.14	3.2
1.35	17.7	1.75	8.0	2.15	3.2

1.36	17.4	1.76	7.8	2.16	3.1
1.37	17.1	1.77	7.7	2.17	3.0
1.38	16.8	1.78	7.5	2.18	2.9
1.39	16.5	1.79	7.3	2.19	2.9
1.40	16.2	1.80	7.2	2.20	2.8
1.41	15.9	1.81	7.0	2.21	2.7
1.42	15.6	1.82	6.9	2.22	2.6
1.43	15.3	1.83	6.7	2.23	2.6
1.44	15.0	1.84	6.6	2.24	2.5
1.45	14.7	1.85	6.4	2.25	2.4
1.46	14.4	1.86	6.3	2.26	2.4
1.47	14.2	1.87	6.1	2.27	2.3
1.48	13.9	1.88	6.0	2.28	2.3
1.49	13.6	1.89	5.9	2.29	2.2
1.50	13.4	1.90	5.7	2.30	2.1
1.51	13.1	1.91	5.6	2.31	2.1
1.52	12.9	1.92	5.5	2.32	2.0
1.53	12.6	1.93	5.4	2.33	2.0
1.54	12.4	1.94	5.2	2.34	1.9
1.55	12.1	1.95	5.1	2.35	1.9
1.56	11.9	1.96	5.0	2.36	1.8
1.57	11.6	1.97	4.9	2.37	1.8
1.58	11.4	1.98	4.8	2.38	1.7
1.59	11.2	1.99	4.7	2.39	1.7
1.60	11.0	2.00	4.6	2.40	1.6

TABLE 13.2(c)
Normal distribution
Percent of area outside specified deviation

Deviation	Percent	Deviation	Percent	Deviation	Percent
2.41	1.6	2.81	.5	3.21	.1
2.42	1.6	2.82	.5	3.22	.1
2.43	1.5	2.83	.5	3.23	.1
2.44	1.5	2.84	.5	3.24	.1
2.45	1.4	2.85	.4	3.25	.1
2.46	1.4	2.86	.4	3.26	.1
2.47	1.4	2.87	.4	3.27	.1
2.48	1.3	2.88	.4	3.28	.1
2.49	1.3	2.89	.4	3.29	.1
2.50	1.2	2.90	.4	3.30	.1
2.51	1.2	2.91	.4	3.31	.1
2.52	1.2	2.92	.4	3.32	.1
2.53	1.1	2.93	.3	3.33	.1
2.54	1.1	2.94	.3	3.34	.1
2.55	1.1	2.95	.3	3.35	.1

2.56	1.0	2.96	.3	3.36	.1
2.57	1.0	2.97	.3	3.37	.1
2.58	1.0	2.98	.3	3.38	.1
2.59	1.0	2.99	.3	3.39	.1
2.60	.9	3.00	.3	3.40	.1
2.61	.9	3.01	.3	3.41	.1
2.62	.9	3.02	.3	3.42	.1
2.63	.9	3.03	.2	3.43	.1
2.64	.8	3.04	.2	3.44	.1
2.65	.8	3.05	.2	3.45	.1
2.66	.8	3.06	.2	3.46	.1
2.67	.8	3.07	.2	3.47	.1
2.68	.7	3.08	.2	3.48	.1
2.69	.7	3.09	.2	3.49	.0
2.70	.7	3.10	.2	3.50	.0
2.71	.7	3.11	.2	3.51	.0
2.72	.7	3.12	.2	3.52	.0
2.73	.6	3.13	.2	3.53	.0
2.74	.6	3.14	.2	3.54	.0
2.75	.6	3.15	.2	3.55	.0
2.76	.6	3.16	.2	3.56	.0
2.77	.6	3.17	.2	3.57	.0
2.78	.5	3.18	.1	3.58	.0
2.79	.5	3.19	.1	3.59	.0
2.80	.5	3.20	.1	3.60	.0

Now, apply this to the public opinion poll where 520 respondents chose candidate A and 480 chose candidate B. Again, if there is equal preference for the two in the sampled population, what is the probability of 520 or more favoring a given candidate?

This is a deviation of 20 over the expected arithmetic mean, and (as you shall see a little later) the standard deviation is 15.8. After you translate this to the normal distribution by dividing 20 by 15.8 to get a normal deviate of 1.27, you see in Table 13.2 that 20.5 percent of a normal distribution lies outside the mean +/− 1.27 standard deviation units.

This, then, is the answer you are looking for. If you worked out all the possible combinations in tossing 1,000 well-balanced coins, the frequencies derived would show the same thing; for heads that deviated from the mean of 500 by 20 or more in either direction, a probability of 20.5 percent.

In this case, where random chance alone will produce a deviation of this magnitude about one-fifth of the time, most analysts would not consider the poll as persuasive evidence that there is a true margin in favor of candidate A in the sampled population.

To find the standard deviation in a binomial population, the simple formula, with s as the standard deviation, n as the number of events, p as the probability of a given occurrence, and q as p-1, is this:

$$s = \sqrt{npq}$$

For the poll data this is the square root of 1,000(.5)(.5), which works out to be 15.8.

STATISTICAL SIGNIFICANCE

The usual practice is to consider a deviation to be statistically significant only if it would occur by random chance less than 5 percent of the time, or—a more rigorous standard—less than 1 percent of the time.

You have already seen that the 5 percent value is 1.96 times and the 1 percent value is 2.58 times the standard deviation.

Since standard deviation in a binomial distribution when p and q are expressed as percentages is $\sqrt{pq/n}$ it is very simple to calculate this value for any specified percentage. If p is 50 percent, for example, and n is 100, the formula for the standard deviation works out as follows:

$$SD = \sqrt{50 \times 50/100} = \sqrt{25} = 5$$

This means that even if the true value of p in the target population is 50 percent, you can expect a sampling result less than 45 or more than 55 percent about one-third of the time purely as a result of random chance.

When you multiply 5 by 1.96 you get 9.8, which means that in a sample of 100 you can expect the percentage to be less than 50 minus 9.8 or greater than 50 plus 9.8 about 5 percent of the time.

Similarly, when you multiply 5 by 2.58 you get 12.9 which means that the percentage in a sample of this size should be less than 50 minus 12.9 or greater than 50 plus 12.9 about 1 percent of the time.

Table 13.3 shows similar values where the size of the sample ranges from 100 to 50,000. Clearly, sampling variability is reduced and accuracy is improved accordingly as the samples grow in size. But this becomes increasingly inefficient because reliability increases with the square root of n. In other words, to cut sampling error in half, you must increase sample size four times!

This has the practical effect of putting rather definite limits on sample size in surveys of all kinds. After a certain point, typically a sample size of one to two thousand, the marginal reduction of sampling error would normally be considered too negligible to justify the cost of a larger sample.

For example, as the table shows at the 2,000 size level, if p in the target population is 50 percent, you can expect the sampling results to fall within the range of about 48 to 52 percent about 95 times in 100, and doubling the sample would reduce this error range only about half a point in either direction.

NULL HYPOTHESIS

Now it is time to review the basic concept in this analysis, with another perspective on statistical significance.

First, set up what is called a *null hypothesis*, which assumes that each candidate is favored by 50 percent of the target population. The election poll of 1,000 respondents is, in effect, a test of that hypothesis.

TABLE 13.3
Sampling variability
P = 50%

Number	Percentage point accuracy	
	.05 Level	.01 Level
100	9.80	12.88
200	6.93	9.10
300	5.66	7.43
400	4.90	6.44
500	4.38	5.76
600	4.00	5.26
700	3.70	4.87
800	3.46	4.55
900	3.27	4.29
1000	3.10	4.07
2000	2.19	2.88
3000	1.79	2.35
4000	1.55	2.04
5000	1.39	1.82
6000	1.27	1.66
7000	1.17	1.54
8000	1.10	1.44
9000	1.03	1.36
10000	.98	1.29
20000	.69	.91
30000	.57	.74
40000	.49	.64
50000	.44	.58

Any poll result is subject to sampling error, but it is possible to set probability limits on this. Fix this at the .05 level (where there is only one chance in twenty that a given result is likely to occur by reason of sampling variation), which is a reasonable standard for this type of survey.

Now, as shown in Table 13.3, this means that any poll result between 50 +/− 3.1 percentage points must be considered as consistent with the null hypothesis of equal preference for the two candidates in the population being sampled. Only if the sample shows more than 53.1 percent preference for a candidate will you reject this hypothesis and conclude, instead, that there is a real margin in favor of that candidate in that population being polled.

In your poll, Candidate A was leading with a 52.0 percent favorable vote of the respondents. Clearly, this is well within the 50 +/− 3.1 range, and thus you must accept the null hypothesis of no real difference between the two candidates in the sample population. In other words, this 2 percent margin over 50 percent is statistically nonsignificant.

This may seem like a long way around to arrive at an answer, but it may help to clarify the basic reasoning involved. Again, the poll is like tossing 1,000 coins and counting the number of heads and tails. If the coins are well balanced, the law of probability gives the odds of getting any number of heads in such a toss and thus the probability of any deviation from the expected mean of 50 percent, purely as a matter of chance. All this is worked out in tables based on the normal distribution.

Those conducting and reporting the poll will likely reduce all this to a simple statement that the sampling error is three percentage points. This alerts the user of the poll to the fact that the two-percentage-point margin of Candidate A is subject to considerable doubt.

This is how it is possible to make inferences from a sample and put probability limits on how likely it is that the inferences are correct. What seemed quite mysterious is really very simple, once the mechanism is made clear.

It is, nevertheless, a most ingenious and useful procedure, which opens all kinds of doors to the acquisition of knowledge. The sampling process is a powerful tool for all who know how to use it, and executives who are inept in its use will always be at a real disadvantage in the business world.

PRACTICAL SURVEY PROBLEMS

By way of postscript, it should be emphasized that this entire analysis has been based on the assumption of *random* sampling—where each member of the target population (or each member of any of its stratified segments) has an equal chance of being selected in the sample. If this is not the case, then the principles outlined above do not apply, and there is no way to measure the reliability of sampling results.

This is the real problem in practical survey work. It is easy enough to set the number of respondents needed in the survey to meet any specified reliability limit, but it is often very difficult to see to it that each potential respondent has an equal chance of being represented in the sample and does, in fact, respond.

So be on guard for these elements, and if a survey fails to meet the proper criteria for sample selection and response, treat its results with appropriate doubt and suspicion! A fuller discussion of all this appears in the chapter on Sampling Techniques.

NORMAL DISTRIBUTION

The normal distribution has far greater use than the single application cited in this chapter. It is the means of testing the probability distribution of many statistics.

Even though the original data may be far from normally distributed, many statistics computed from the data, such as the arithmetic mean, do tend to be normally distributed. A mathematical law known as the central limit theorem gives the normal distribution its great power and generality.

A number of additional applications will appear in the chapters that follow.

SUMMARY

It is important to understand the logic in the process of making statistical inferences from sampling data. It is an ingenious application of the laws of probability, leading to many techniques of great practical utility. All this can open many new doors to knowledge and answer many questions that would otherwise remain a mystery.

CHAPTER 14
Interpreting Survey Results

Suppose you are the marketing manger for a food product, and a survey company, based on diaries kept by 100 consumers of the product, reports that these households consume an average of 12.4 packages of the product per annum. How much faith can you put in this reported average?

If you could conduct further surveys of this kind and compute the arithmetic mean in each case, then you could compute the standard deviation of these means as a measure of the variability to be expected from the sampling process. That, of course, is impractical, but, happily, there is a quick and easy means of estimating what this variability would be if you did repeat the survey procedure over and over again.

On the assumption that these means are normally distributed, which usually is the case, it can be established mathematically that this measure of variability among sample means (which is called the *standard error* of the mean) is the standard deviation of the sample divided by the square root of the number in the sample. And this can be interpreted like any other deviation in the table on the normal distribution.

The survey company will know the standard deviation of usage among the various households in the sample, and assuming this to be 8.4, the standard error of the mean then works out as follows:

$$S_M = \frac{\text{Standard Deviation}}{\sqrt{N}}$$

$$S_M = \frac{8.4}{\sqrt{100}} = .84$$

This you can now interpret as you would any other standard deviation in a normal distribution. In other words, the odds are about two out of three that the true mean lies between 12.4 +/− .84. And the odds are about 95 in a 100 that the true mean is within the range of the sample mean and twice the standard error, i.e., 12.4 +/− 1.68.

And so, with this little bit of magic, you arrive at a very precise measure of the confidence that can be placed in the reported survey results. It is not necessary to repeat the sampling process again and again, computing the mean each time, to measure sampling variability. The shortcut formula, in effect, accomplishes the same thing from nothing more than data within the sample itself.

The formula makes good sense, of course. The greater the variability within each sample, the greater the variation to be expected in the means of repeated samples. And the larger the sample size, the less the variability to be expected in such means.

DIFFERENCE BETWEEN TWO PERCENTAGES

Take another example. Once again you are the executive in charge of marketing a new food product and must choose between two package designs. In a controlled test against competitive brands, package A was chosen by 18 percent of the respondents, and package B by 15 percent. A total of 200 consumers participated in the test.

The question to you: how much reliance can you place in this survey finding? Is the three-percentage-point difference of any real significance, or could it easily arise from random sampling variation?

As in the first example, if you had repeated tests of this kind and knew the difference between package acceptance in each test, you would have a direct measure of the sampling variability to be expected, but of course this is quite impractical. Once again, however, there is a short-cut method of estimating this variability from data in the sample itself.

Again, if these differences are normally distributed, the procedure is to square the standard error of the mean for each category of the sample, add the two squares, and take the square root. The result is the *standard error of the difference* between the two means involved, which, once again, can be interpreted like any other deviation in the normal distribution.

This is further simplified in dealing with percentages since the standard error of a percentage is simply $\sqrt{pq/n}$ where p is given percentage, q is 100-p, and n is the number in the sample. Thus, the standard error squared is simply pq/n.

In the above example, therefore, the standard error of the difference between the two percentages works out as follows:

$$S_d^2 = pq/n + pq/n$$

$$= \frac{(18)(82)}{200} + \frac{(15)(85)}{200}$$

$$S_d^2 = 13.755$$

$$S_d = 3.71$$

And this is the answer you are looking for. In repeated tests of this kind, you would expect the difference between the two reported percentages to vary with a standard deviation (or standard error) of 3.71 percentage points. Knowing this, you can evaluate the significance of the reported difference in package preference in the controlled test.

For example, you can set up the null hypothesis that there is no real difference in package preference, with the reported difference being due wholly to sampling variation, and test the probability of this.

If the true difference is zero, then the reported difference of three percentage points is equivalent to a difference of 3/3.71 or .81 in standard deviation units. Reference to the normal table shows that 41.8 percent of the area in a normal distribution lies outside a deviation of this size.

In other words, in repeated tests of this kind, even if the true difference is zero, you would expect to get a difference of three or more percentage points about 42 percent of the time. Thus, in terms of decision making, the given survey result leaves much to be desired. For all practical purposes, it is not much better than flipping a coin to make a choice between the two packages.

What you do at this point, of course, depends on other considerations. If you must make an immediate decision and the two designs are equal in cost and other key factors, then it would make sense to choose Package A even though you have a very low confidence level in its relative superiority.

But if Package A is more costly than B, or is less preferable on other grounds, and you have more time to make the decision,

104 BASIC BUSINESS STATISTICS FOR MANAGERS

then you may wish to expand the original survey with additional testing.

In any event, this procedure has given you a very precise knowledge of the true odds involved in the apparent superiority of Package A over B, and your decision can be made accordingly. This is very different from a decision based on nothing more than blind faith in the survey results.

DIFFERENCE BETWEEN TWO MEANS

Now, try a different question, similar in nature but not involving percentages.

Once again, you are a marketing manager and need to choose between two versions of your product: A or B. Product A has been supplied to 100 consumers, and Product B to another 100 consumers. Product usage has been measured for each consumer, and the average use per month has been 4.7 boxes of Product A and 3.8 boxes of Product B.

The question to you: is this sufficient evidence to choose A over B? And to answer this, you must once again ascertain the probability that random chance alone might have produced the reported difference in usage.

For this you again need the standard error of the difference, and you will recall the formula for this. First you must compute the standard deviation of usage among all participants in each sample and divide this by the square root of the number to get the standard error of the mean in each case.

Then square and sum the two standard errors, and take the square root to get the standard error of the difference. The survey company provides the standard deviation within each group, and all other calculations are shown below:

	A	B	Sum
Mean..................	4.7	3.6	
Standard deviation	2.8	2.4	
N (number)..........	100	100	
St. error of mean	.28	.24	
Squared............	.0784	.0576	.1360
St. error difference			.3688

Now, to test the null hypothesis that there is no real difference in usage of the two products, you must consider how likely it is that a difference of 1.1 would be reported, which is equal to 3.0 standard deviation units (1.1/.3688). A glance at the normal table shows that there is far less than one chance in a 100 that such a deviation would occur purely as a matter of chance.

Thus you would conclude that Product A is indeed superior to B in terms of household consumption, and further conclude that the odds are about two to one that the real difference is within the range of 1.1 plus or minus the standard error, i.e., 1.1 +/− 0.37.

PAIRED SAMPLES

Now, consider another example. You are testing two new machines and must decide which to purchase. Five operators use machine A, and another five Machine B. Output of each operator is shown below:

	A	B
	12	11
	8	20
	14	15
	10	9
	18	14
Average	12.4	13.8

Operators using Machine B averaged 1.4 more output units than those using Machine A, but how much reliance can you place in this difference in making your decision?

Using the procedure described above for testing the difference between two means, you find that the standard error of the difference in this case is 2.59. The survey difference of 1.4 is equal to only .54 standard deviation units, and the normal table shows that a deviation this large or larger will occur due to random chance alone about 59 percent of the time.

Clearly, these data are quite insufficient to establish any real difference between the two machines. It would take a much larger sample for such a difference to be statistically significant.

But suppose that in setting up this test you decided to use the same five operators to test both machines, and you have the output for each operator, as follows:

Operator	A	B	Difference
1	10	11	+1
2	18	20	+2
3	12	14	+2
4	14	15	+1
5	8	9	+1
Average ...	12.4	13.8	+1.4
Standard error			.6

Now look at the difference column only. The standard deviation of the numbers in this column is .6, and the difference of 1.4 is equal to 2.3 standard deviation units. As you know from the normal distribution, the chance of getting a deviation as large or larger than this is less than 1 in 100. Thus you would conclude in this case that the difference in output between the two machines is highly significant from a statistical viewpoint.

You may wonder, at this point, how this can be. The data are the same, it would appear, but the conclusion as to statistical significance is altogether different.

In fact, the data are not the same. In the first instance, since there were different operators, there were two separate samples. In the second instance, you are dealing with only one sample of five operators. In effect, all the variation between operators has been excluded, and this adds enormously to the precision of the test.

In setting up tests and surveys of all kinds, this is a most important principle to keep in mind. In testing various alternatives, the use of the same participants has the effect of eliminating variation from one to another and thus is almost certain to be far more efficient than selecting different participants for each such test.

SUMMARY

Competent survey companies will use the procedures explained in this chapter to evaluate the significance of their reported results and will make such data available to their client. If you are familiar with these procedures, you can understand and fully appreciate what is being reported.

If, on the other hand, whoever conducts the survey fails to specify the statistical significance of reported results, you should insist that these tests be made before placing any real reliance in the survey findings.

CHAPTER 15
Sampling Techniques

The procedures described in this book for testing the reliability of sampling data are based on the assumption that the sample was chosen by a random process, and if this assumption is not correct, then the procedures are no longer valid.

The term *random* is used here in a technical sense. It does not mean casual or happenstance. It means that each member of the target population, or defined segment thereof, has an equal chance of being selected. This is why samples drawn in this fashion, where the probability of selection is part of the sample design, are known as *probability* samples.

It is only through samples of this type that inferences can be made about the target population as a whole, with confidence limits based on the laws of probability.

It is easy enough to define what a probability, or random, sample should be in theory, but it is often quite difficult in practice to meet the necessary criteria.

As an illustration of this, assume there are 1,000 employees in your company and you want to study the personnel records of a sample of 100 of these. How would you select this sample?

It would clearly be wrong simply to pick the 100 names by happenstance. There is no guarantee in such a process that each employee would have an equal chance of being selected.

One way to overcome this is to number the employees from 1 to 1,000 and then select 100 of them using an equal number of random numbers taken from a table of such numbers or from a similar list generated by computer.

This is a pure random sample, and it meets the basic criterion: each employee has one chance in ten of being included in the sample. But there may be a more efficient way to draw such a sample.

For example, employees could be listed alphabetically by department. A random number is drawn from one to ten and the employee in this rank position on the list is chosen along with every tenth individual thereafter.

Here again each employee has an equal chance of being selected, but this method ensures proportionate representation of each department in the sample: clearly an improvement over the pure random system.

STRATIFIED SEGMENTS

Similarly, it is often more efficient to divide a target population into different categories, with random sampling in each category.

For example, in a customer analysis it may be efficient to classify the customers by size. Where the statistic being analyzed is more variable for large than for small customers, the sampling ratio for the former might be 50 percent versus 10 percent for the latter. To combine the two sub-samples, you would then multiply results for the first by two and those for the second by ten for a properly weighted total.

Stratified sampling of this kind is always more efficient when the different categories or strata differ in variability. In general, the greater the variability in a given category, the higher the sampling ratio should be.

To illustrate this with an extreme case, suppose all members of a given category are identical in the characteristic being measured. In such a case, clearly, only one member need be selected to tell you all you want to know about the group as a whole.

Let me emphasize again that none of this violates the basic criterion of random sampling. Each member in the same category has an equal chance of selection; and the results are then weighted to adjust for the different sampling rates in one category versus another.

TELEPHONE SURVEYS

The same principles apply to public opinion polls, market research surveys, and the like, but they can sometimes be quite difficult to achieve in actual practice.

Oddly enough, it is very easy to select a random sample of telephone numbers, whether you are sampling one metropolitan area or the entire United States. Computers can now select and dial such numbers, including unlisted numbers, with great efficiency.

With all such surveys, of course, the results can be projected only to households or individuals with a telephone. They reveal nothing about all other members of the population. But the telephone survey is still a widely used technique because households with a telephone represent a very large percentage of all households in the United States.

CLUSTER SAMPLING

All of this becomes much more difficult in surveys by personal interview. It would be highly inefficient, for example, to select households in a large metropolitan area—much less the nation as a whole—by pure random sampling, which would require a great deal of travel time just to reach the various households selected.

The usual stratagem to overcome this is to divide the area into various geographical sections with clusters of blocks in each section and to select these clusters by random sampling. Then, through some random method, households are selected in each such cluster for inclusion in the sample.

Government agencies and private companies that conduct surveys of this type on a regular basis have worked out all this methodology in great detail in such a way as to minimize interviewer time and travel. Personal interviewing is a costly process as best, and it would be prohibitive in most cases if these methods were not utilized.

NONRESPONSE RATE

A major problem in survey work is the nonresponse rate. The selected respondent may not be at home at the time of the interview, or may refuse or be unable to participate in the survey. The net effect of this can be very serious indeed.

For example, assume a 40 percent nonresponse rate in a given survey. If the survey work goes no further than this, then the results reveal nothing about the 40 percent of the population represented by this nonresponse group. To assume they are like the response group in the characteristics being measured is quite erroneous. Indeed, the precise opposite is more likely to be the case. It is very likely that this group will differ from the response group, but any inference about this is still guesswork, pure and simple.

Thus, survey companies work very hard to reduce the nonresponse rate by using advance calls to set up time for interviews, monetary inducements of one kind or another, and other methods such as expensive recalls.

One stratagem sometimes used is to find out for each respondent the number of days when he or she would have been available for interviewing in the prior week and then to weight all responses in inverse ratio to the number of days thus reported. This is an ingenious and inexpensive method of coping with the out-of-home component of the nonresponse group—not perfect, but perhaps a good working solution nevertheless.

MAIL SURVEYS

One often sees a report based on some mail survey where the sponsor proudly announces a "very high rate of return," such as 30 percent, as proof of the survey's validity. This, of course, is nonsense of the first order. What the sponsor should say is, "only 30 percent of those on our list returned the questionnaire and we know nothing whatsoever about the other 70 percent and the people they represent!"

An amusing version of this can be seen in mail surveys of those who graduated from some college in some prior year. Results usually indicate that most of the graduates have achieved a high degree of success with compensation to match. It takes no great imagination to infer that those of lesser achievement, often a large percentage of the total, had no desire to participate in the survey and simply dropped the questionnaire in the nearest wastepaper basket!

It is possible to correct this deficiency in mail surveys by treating the nonresponse group as a separate category and by having follow-up sampling within the group by interviews in person or by telephone. Results can then be weighted by the numbers in each group to get combined totals for the entire sample.

SIZE OF SAMPLE

You may have been surprised, in the description of techniques used to set error limits on various sampling statistics, by the fact that no reference was made to the size of the population being sampled. The formulas were based on variability within the sample and size of the sample, with no consideration of how large the sample might be as a percentage of its universe.

In other words, a given sample of a certain size and variability would have the same error limits whether it related to a single metropolitan area or to the United States as a whole.

This seems to go against all common sense, but it is nevertheless true for all practical purposes. Theoretically, there is some reduction in sampling error if the sample is a large percentage of its universe, but this happens so rarely that it may as well be ignored. The reason for this is found in the laws of probability that form the basis of all these techniques for analyzing sampling data.

For example, consider a coin-tossing experiment. You flip a coin 1,000 times and find that heads come up 50.2 percent of the time. This sample is drawn from an infinite universe because there is no theoretical limit to how many times you can flip the coin. In such a case, clearly, the size of the sample as a percent of its universe is without meaning. Yet, how many more times would you want to flip this coin to be reasonably certain that heads will come up about half the time?

The obvious answer to this question is verified by the mathematics of probability, which show that there is very little to be gained in this instance by increasing the sample size. You have already learned as much from a practical viewpoint as you need to know about this particular phenomenon. And, of course, exactly the same conclusion applies to polls and surveys where the size of the universe being sampled is equally irrelevant in terms of sampling error.

The practical problem in surveys of this type is not in the size of the sample, which is easy to deal with, but rather in establishment of the mechanism to meet the strict requirements of probability sampling. And clearly this tends to become more difficult with increases in the size of the population and the geographical area to be covered.

SUMMARY

Sampling is an efficient way to answer many questions and, indeed, sometimes the only practical way. A steel beam, for instance, must be bent to test its bearing strength, but this destroys the beam. If sampling were not used and the entire universe were tested, there would be no product left for the steel company to sell!

To be able to set confidence limits on inferences about a target population based wholly on a sample drawn from that population is an ingenious application of the mathematical laws of probability. It is a powerful process with great practical utility, opening all kinds of doors to the acquisition of knowledge.

But the process requires that the sample be drawn in a very precise way so as to meet the strict criteria outlined in this chapter. Although these procedures are well known to professionals in survey work, they are often ignored by survey sponsors who are unwilling to spend the money to do the job correctly. In all such cases, you should treat the survey results with appropriate caution and suspicion.

CHAPTER 16
Chi Square

When reading survey reports, you are likely to encounter something called *chi square,* and you should understand what it means and how it is used. In essence, it is a quick means of testing reported differences by respondent category for relative significance.

To illustrate this, again assume that you are the marketing manager of a food product and must choose between two versions of the product. In a controlled test, those favoring one or the other is as follows:

	Favoring		
	A	B	Total
Men...	62	85	147
Women...	141	127	268
Total...	203	212	415

The question to you: should the reported preference of men versus that of women be considered significant, or could it easily arise from sampling fluctuation even if there is no real difference in the population being sampled? The chi square test can give you a quick answer to this question.

The distribution of chi square is known and the details are shown in published tables. From this you can find the exact probability of exceeding any specified chi square value purely as a matter of chance.

You will see the method of calculation a little later, but for the data as shown, the value of chi square is 4.14, which can be compared with these two values from the chi square distribution:

Chi square	
.05 level...............	3.84
.01 level...............	6.64

These are the values of chi square to be expected 5 percent of the time and 1 percent of the time respectively, purely as a matter of sampling variation, even though there is no real difference between the two categories in the population being sampled.

Thus you see that the chi square value of 4.14 would occur less often than 5 percent of the time, but more often than 1 percent of the time due wholly to random chance.

For most practical purposes, the .05 significance level is considered quite adequate in tests of this kind, and on the basis of this, you would conclude that there is a real difference in the preference of men and women for versions A and B of the product.

CALCULATING CHI SQUARE

It takes a little arithmetic, but it is really quite easy to calculate the value of chi square.

In the case cited above, for example, you are testing the null hypothesis that there is no real difference between men and women in their preference for the two product versions. For the sample as a whole, 48.9 percent preferred Product A and 51.1 percent preferred Product B, and under the null hypothesis you would expect the same percentages for men and women in the sample.

If this were so, then 71.9 men would have chosen A and 75.1 would have chosen B; and similarly, 121.1 women would have chosen A and 136.9 would have chosen B.

Now, you compute chi square by finding the difference between these theoretical numbers and the actual numbers, squaring the difference in each case and dividing by the theoretical number, and then adding the result for each category. If T is the theoretical number and A the actual number, then the formula for chi square is simply:

$$\text{Chi square} = \text{Sum}\ \frac{(T - A)^2}{T}$$

The actual calculations for the example above are shown in Table 16.1, and they lead to the 4.14 result already specified. To test its significance, turn to Table 16.2, which shows chi square values at the .05 and .01 probability levels: values to be expected, due to random chance, 5 percent and 1 percent of the time, respectively.

TABLE 16.1
Data

Category	Group A	B	Total
1	62	85	147
2	141	127	268
Total	203	212	415

Chi square calculation

Group	Theo.	Actual	Diff.	Diff.^{2}	Chi sq.
A	71.9	62	−9.9	98.13	1.36
B	75.1	85	9.9	98.13	1.31
A	131.1	141	9.9	98.13	0.75
B	136.9	127	−9.9	98.13	0.72
Chi square					4.14

Note: computer calculated prior to rounding.

The table involves a concept that has been mentioned before, *degrees of freedom*, which can be illustrated with this example. Although there are four cells in our data table, men and women vs. A and B, there is only one degree of freedom. The reason for this, if all the subtotals and the grand total are fixed, is that a change in any one cell automatically changes the numbers in the other three cells.

The table gives the two probability values of chi square from 1 to 30 degrees of freedom, which covers most of the situations encountered in practice. But beyond this, the formula in the table can be used to convert the chi square value into a normal deviate, which can then be interpreted accordingly.

ANOTHER EXAMPLE

Now that the arithmetic is clear, try the chi square test in another situation. Once again, assume a controlled test of Product A versus

TABLE 16.2
Chi square

Degrees of freedom	Level	
	.05	.01
1	3.84	6.64
2	5.99	9.10
3	7.82	11.34
4	9.49	13.28
5	11.07	15.09
6	12.59	16.81
7	14.07	18.48
8	15.50	20.09
9	16.92	21.67
10	18.31	23.21
11	19.68	24.73
12	21.03	26.22
13	22.36	27.67
14	23.69	29.14
15	25.00	30.58
16	26.30	32.00
17	27.59	33.41
18	28.87	34.81
19	30.14	36.19
20	31.41	37.57
21	32.67	38.93
22	33.92	40.29
23	35.17	41.64
24	36.42	42.98
25	37.65	44.31
26	38.89	45.64
27	40.11	46.96
28	41.34	48.28
29	42.56	49.59
30	43.77	50.89

For n greater than 30 the value: $T = \sqrt{2X^2} - \sqrt{2n-1}$ can be interpreted as a normal deviate.

others in households selected to represent various market regions, with these results:

Region	Total	Favor A	Percent
1	128	41	32.0
2	113	33	29.2
3	142	42	36.6
4	187	55	29.4
5	135	48	35.6
6	192	60	31.3
Total	897	289	32.2

Details of the calculation are shown in Table 16.3, and the resulting chi square value is 3.18. There is one degree of freedom in each region and thus six altogether, and from Table 16.2 you see that this is far short of the 12.59 value at the .05 probability level.

On the basis of this, it is evident that the reported regional differences in the sample could easily arise from random chance alone and thus have no real statistical significance.

SUMMARY

The chi square procedure is a useful, quick test of the statistical significance of many survey findings. The simple arithmetic involved is a small price to pay for the knowledge gained, and all executives who understand this procedure can benefit accordingly.

**TABLE 16.3
Data**

Category	Group A	B		Total
	41	87	1	128
	33	80	2	113
	52	90	3	142
	55	132	4	187
	48	87	5	135
	60	132	6	192
	289	608	Total	897

Chi square calculation

Group	Theo.	Actual	Diff.	Diff.^{2}	Chi sq.
A	41.2	41	−0.2	0.06	0.00
B	86.8	87	0.2	0.06	0.00
A	36.4	33	−3.4	11.61	0.32
B	76.6	80	3.4	11.61	0.15
A	45.8	52	6.2	39.06	0.85
B	96.2	90	−6.2	39.06	0.41
A	60.2	55	−5.2	27.55	0.46
B	126.8	132	5.2	27.55	0.22
A	43.5	48	4.5	20.30	0.47
B	91.5	87	−4.5	20.30	0.22
A	61.9	60	−1.9	3.46	0.06
B	130.1	132	−1.9	3.46	0.03
	Chi square 3.18				

Note: computer calculated prior to rounding.

CHAPTER 17
Analysis of Variance

A great deal of scientific knowledge has been derived from experiments of various kinds. There is less opportunity for this in the business world, but such opportunities do exist and executives should be aware of their potential.

They should also be familiar with at least the rudiments of good experimental design and know how to analyze results. One important analytical procedure is called *analysis of variance* or, in abbreviated form, *ANOVA*. This chapter will try to explain the concept and basic elements of ANOVA.

This approach was developed primarily for the design and analysis of experiments in the field of agriculture in an effort to measure the effect on crop yield of various elements such as fertilizer, type of seed, and so on.

The problem here is that no two plots of land, no matter how close and apparently similar, are really the same, and there is no simple way to control or exclude the effect of this in experimental results. A tour of the wine country in France is graphic proof of this. Wine produced by one small estate can be vastly different from that produced by others in the immediate neighborhood.

The solution that emerged was to select the experimental plots for each treatment on a random sampling basis and then to analyze the experimental results in terms of error to be expected from random sampling. The variation could not be controlled, but at least it could be measured and analyzed in probability terms. This is the essence of ANOVA.

The best way to understand this procedure is by way of an example. Assume that you are the advertising manager of a direct mail company and want to check the relative response to three different mailings to five separate mailing lists.

You take 1500 sample names from each list, divide them at random among the three mailings, send them all out and then, after an appropriate time, check the return rate of each. Results are shown in the upper half of Table 17.1.

TABLE 17.1
Return per 1000

List	A	B	C	Total	Average
1	190	186	202	578	192.7
2	169	174	203	546	182.0
3	189	218	217	624	208.0
4	197	217	223	637	212.3
5	172	204	189	565	188.3
Total....	917	999	1034	2950	
Average..	183.4	199.8	206.8		196.7

Source of variation	Deg of freedom	Sum of squares	Mean square	F
Rows	4	2023	505.84	5.09
Columns ..	2	1443	721.31	7.26
Remainder .	8	795	99.42	
Total	14	4261	304.38	

Note: computer calculated prior to rounding.

Clearly, the return rate differs by mailing and by list, and the question is whether these differences are significant or whether they could easily arise from random chance.

SUM OF SQUARES

The first step is to measure the total *variance* in the sample, which is the deviation of each cell value from the average for the entire sample, squared and added. For this, you can use the short-cut formula described in the chapter on *Standard Deviation*, which works with the original numbers without the need for finding the deviation from the mean in each individual case.

You will recall that this formula, with X being the original number in each case and N the total number of items, is:

$$\text{Sum of Squares} = \text{Sum } (X)^2 - (\text{Sum } X)^2/N$$

122 BASIC BUSINESS STATISTICS FOR MANAGERS

When this is applied to the direct mail sample, it gives:

Sum of Squares $= 584,428 - (2,950)^2/15 = 4,261$

Now, if this is the total variation in the sample, the next question is how this is broken down into its components: how much is due to the variation from one mailing to another, how much to the variation from one list to another, and the residual after these two have been subtracted.

To find the sum of squares associated with the different mailings, take the average for each mailing as a deviation from the overall average, square this difference and multiply by the number of cells involved, and sum these totals, as follows:

Mailing	Avg.	Dev.	Dev.²	x 5
A	183.4	13.3	176.89	884.48
B	199.8	3.1	9.61	48.05
C	206.8	10.1	102.01	510.05
Sum				1442.58

Following the same procedure with the row averages (where the number of cells involved is three), you get:

List	Avg.	Dev.	Dev.²	x 3
1	192.7	4.0	16.00	48.00
2	182.0	14.7	216.09	648.27
3	208.0	11.3	127.69	383.07
4	212.3	15.6	243.36	730.08
5	188.3	8.4	70.56	211.68
Sum				2021.10

This is the sum of squares associated with the different mailing lists used. (Because the average is carried to only one decimal point, there is a small rounding error, and the true value is 2023.)

The remaining sum of squares in the sample is the total of 4,261 less the 1,443 associated with the different mailings and less the 2,023 associated with the various mailing lists, or a net of 795.

You would get the same value for the remaining sum of squares if you calculated the deviation of each cell from the mean of the row and column averages for that cell, then squared the deviations, and summed the squares. Clearly then, it represents all of the variance in the sample not associated with the two control elements.

TESTING F VALUES

All of this is recapped in Table 17.1. The degrees of freedom are one less than the number involved in each case. When the sum of squares is divided by the degrees of freedom, you have the *mean square* for each component. The F value is derived by dividing the given mean square by the mean square for the remainder.

Remember that the mean square of the remainder is a measure of random sampling variation, and thus the F value shows the relative variance in each control element versus the random component in the sample. Clearly, the larger the value of F the more likely it is to be statistically significant, but to gauge this precisely you need to refer to published tables.

An abbreviated version of this appears in Table 17.2, where you see the F values for different degrees of freedom at the .05 level of

TABLE 17.2
Table of F values
.05 Probability Level

Smaller D.F.	Larger degree of freedom							
	1	2	3	4	5	10	100	500
2	18.51	19.00	19.16	19.25	19.30	19.39	19.49	19.50
3	10.13	9.55	9.28	9.12	9.01	8.78	8.56	8.54
4	7.71	6.94	6.59	6.39	6.26	5.96	5.66	5.64
5	6.61	5.79	5.41	5.19	5.05	4.74	4.40	4.37
6	5.99	5.14	4.76	4.53	4.39	4.06	3.71	3.68
7	5.59	4.74	4.35	4.12	3.97	3.63	3.28	3.24
8	5.32	4.46	4.07	3.84	3.69	3.34	2.98	2.94
9	5.12	4.26	3.86	3.63	3.48	3.13	2.76	2.72
10	4.96	4.10	3.71	3.48	3.33	3.97	2.59	2.55
20	4.35	3.49	3.10	2.87	2.71	2.35	1.90	1.85
100	3.94	3.09	2.70	2.46	2.30	1.92	1.39	1.30
1000	3.85	3.00	2.61	2.38	2.22	1.84	1.26	1.13

probability. In other words, there is less than one chance in 20 that any F value beyond that shown in the table would occur as a result of random sampling alone.

In the case of the different mailings, you have an F value of 5.09, which compares with the value of 4.46 in the table; and in the case of the different lists, the F value of 7.26 compares with the value of 3.84 in the table. Thus you see that there is much less than one chance in twenty that these F values were generated by sampling variation alone.

The conclusion is that the sample differences by type of mailing, and by mailing list, are significant, and decisions can be based on the data accordingly.

SUMMARY

It clearly was very efficient in this case to test three different mailings and five different lists in the same experiment, and the ANOVA procedure tells you whether or not the results are statistically significant.

The arithmetic in the analysis of variance may seem forbidding at first, but it really is quite simple and straightforward. The key fact is that variance is additive and thus can be broken down into its several components for analysis of each.

Needless to say, this discussion has barely touched the subject of experimental design, on which entire books have been written, but I do hope it has given you some grasp of the concept and how the results can be interpreted: a powerful tool in the pursuit of scientific knowledge.

CHAPTER 18
Regression and Correlation

In the business world, as in many other fields, the acquisition of knowledge often depends on understanding the relationship between two sets of numbers, and executives should be familiar with the techniques used to unravel this mystery in a procedure called *regression analysis*.

This rather odd name is traceable to research many years ago by an eminent British scientist named Galton, who studied the relationship between the heights of sons and their fathers. He found that sons of tall fathers tended to be taller than average but not as tall as their fathers, and those with short fathers tended to be shorter than average but not as short as their fathers. He described this process as *regressing* to the racial average, and thus it is that *regression analysis* is still used to describe the procedure for measuring the relationship of one variable to another.

It is a good thing, incidentally, that this law of regression exists, because otherwise a few generations would be sufficient to split the human race into two species: one of dwarfs and the other of giants!

Very likely the law operates in many other fields as well. The star athletic performers one year are likely to be above average the next year, but not likely, in general, to equal or exceed their performance of the prior year. And the same is true of mutual funds, stocks, and a wide range of other phenomena.

In any event, since it is important to understand how to analyze all these things, you should begin with the simplest kind of relationship, which can be represented by a *linear equation:*

$$Y = a + bX$$

Here Y and X are variables and a and b are fixed values called *coefficients*. Y is the *dependent* variable because its value depends on the value assigned to X, which is called the *independent* variable.

As an illustration, suppose on certain checking accounts a bank imposes a service charge of $5 per month plus $.10 per check. With

Y as the service charge and X as the number of checks, this can be written:

$$Y = 5 + .10X$$

Clearly, Y shows the monthly service charge for any specified number of checks (X) written that month.

In charting data of this kind (see Figure 18.1), it is customary to use the vertical scale for the dependent variable and the horizontal scale for the independent variable.

The coefficient a is called the *intercept* because it is the value of Y when X is zero and thus represents that intercept value. The coefficient b is the slope of the line: the amount by which Y increases for

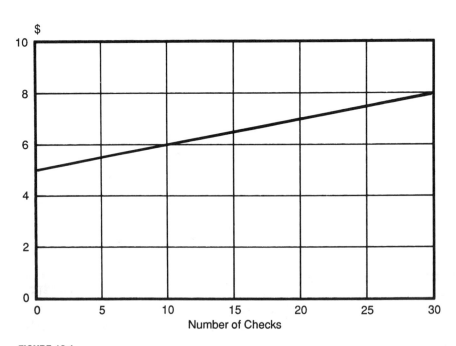

FIGURE 18.1

every unit increase in X. Thus, in the above equation, the coefficient b means that the charge goes up ten cents for each check written.

Any linear equation, when plotted on an arithmetic chart, is a straight line, and thus you need to calculate only two points to draw the line. In this case, you could pick the $5 charge for zero checks and the $8 charge for thirty checks, and the line connecting these two points represents the entire relationship for zero to thirty checks per month.

On the inference that the bank is simply recovering its costs with this service charge, the equation indicates that it costs the bank $5 a month simply to maintain this type of checking account and an additional 10 cents to process each check.

The linear equation is a very useful device and can be used to describe a great many relationships between two sets of numbers. There is no problem in deriving such an equation in the above example because the coefficients a and b are known precisely. But in most relationships, these coefficients are not known and must be estimated, and the discussion of the technique for this follows.

METHOD OF ANALYSIS

Assume that you are the manager of a plant producing a single product and are interested in knowing whether your expenses are under proper control. Your problem is that production varies from month to month, depending on sales orders, and your expenses naturally tend to vary with volume of production, but you do not understand the precise relationship involved.

This is not unusual. While cost accounting data is typically available in great detail, it is still impossible to break down many items into fixed versus variable expenses because they are some composite of both; and unless some way can be found to make this separation, there is no way to understand how total cost should vary with volume of production.

Assume the following total cost (in $000) and total output (in 000 units) for the last six months. In reality, of course, more months should be included in an analysis of this kind, but these limited data will simply illustrate the methodology.

128 BASIC BUSINESS STATISTICS FOR MANAGERS

Month	Production	Cost
1	122	$711
2	146	836
3	113	673
4	125	732
5	116	697
6	137	785

Now, how do you unravel this relationship and, with Y representing cost and X representing production, find the coefficients a and b that best fit the data?

If you plot the data (see Figure 18.2), you see that the various points seem to fall along a straight line, and you could simply draw the line that seems to fit best. The problem with this procedure, of

TOTAL PLANT COST PER MONTH
In Relation to Production

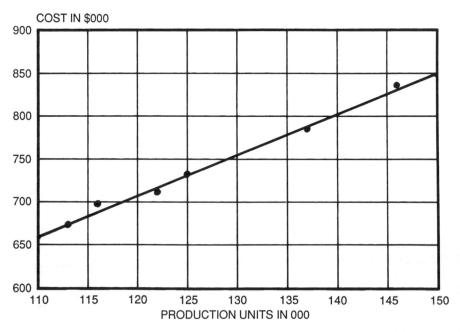

FIGURE 18.2

course, is that it is subjective, and someone else might draw a very different line. The question is how do we determine the one line that, on an objective basis, is the best line of fit?

The method of accomplishing this goes by the rather odd name of *least squares* solution, which specifies the linear equation where deviations from the fitted line, when squared and summed, are a minimum.

The deviations are squared for the same reason they are in computing the standard deviation. If deviations are calculated from *any* straight line drawn through the point representing the average for the two variables, the sum is always zero. So it is necessary to square the deviations to get values that can be algebraically summed.

What you are looking for are the coefficients in the equation that will yield estimated values as close as possible to actual values, as measured by the sum of squared deviations between the two. The arithmetic involved in shown in Table 18.1.

That table shows the actual values of X and Y designated by capital letters; then the deviation of each item from its mean, designated by the small letters x and y. Next come the square of x, the square of y, the product xy, and, finally, the sum of all these values. Using these data, the two coefficients are then calculated by the formulas shown in the table, so that the desired equation is:

$$Y = 135.47 + 4.77X$$

This means that the plant cost each month is composed of a fixed component of $135,470 plus a marginal, or incremental, cost of $4.77 for each unit produced. Now, you need to know how accurate this equation is in estimating cost, and this is shown in the lower half of the table, where you see the estimated and actual value, the deviation between the two, and the square of this deviation.

You will note that the sum of deviations of actual from estimated Y is zero, and this will always be true when the equation is derived by least squares. You may also be assured that the sum of the squared deviations, or 143 in this case, is a minimum. No other straight line can be drawn that will produce a smaller sum.

It is obvious from the table of estimated values, indicated by the straight line in Figure 18.2, that the equation provides an excellent fit to the actual data, which means that you can rely on it as a proper relationship of cost to volume of production.

TABLE 18.1
**Fitting a linear equation
By least squares
Y = a + bX**

X	Y	x	y	x²	y²	xy
122	711	− 5	− 28	20	784	126
146	836	20	97	380	9409	1892
113	673	− 14	− 66	182	4356	891
125	732	− 2	− 7	2	49	11
116	697	− 11	− 42	110	1764	441
137	785	11	46	110	2116	483
759	4434	0	0	806	18478	3843

$$b = \text{Sum}(xy) / \text{Sum}(x2) = 4.77095$$
$$A = [\text{Sum}(Y) - b\text{Sum}(X)] / N = 135.4749$$

X	Est. Y	Y	Dev.	Dev.²
122.00	717.53	711.00	− 6.53	42.65
146.00	832.03	836.00	3.97	15.73
113.00	674.59	673.00	− 1.59	2.53
125.00	731.84	732.00	0.16	0.02
116.00	688.91	697.00	8.09	65.53
137.00	789.10	785.00	− 4.09	16.77
759.00	4434.00	4434.00	0.00	143.24

And so, you as the plant manager, with this little bit of arithmetic, have learned a good deal more about the cost of your operation than you could have learned from weeks of poring over detailed cost accounting data. With this equation, you can estimate what your cost should be in the future based on actual production and thus see precisely how far you miss the target each month in the future.

In actual practice, of course, you need more months in your sample than are shown here, and you may need to make some adjustment for changes in wage rate and the like; but the basic concept and analytical procedure is the same.

ADDITIONAL APPLICATIONS

Some other interesting statistics can be derived from the least squares procedure, as summarized below:

Standard error of estimate: the square root of the sum of squared deviations from estimated Y, divided by the degrees of freedom involved. In the case of a linear equation, the latter is the number of pairs minus two, because the two coefficients take up two degrees of freedom. In other words, a linear equation will fit two points perfectly with no deviation; and with three points, only one degree of freedom exists, and so on.

In the above example, since the sum of the squared deviations is 143.24 and the degrees of freedom is four, the standard error of estimate is the square root of 143.24/4, which is equal to 5.98. This can be interpreted like any other normal deviate. In other words, the chances are about two out of three that any estimate made from the equation will not deviate from the actual value by more than 5.98— and, of course, one chance in three that it will deviate by more than this.

A glance at the deviations in the table shows that this worked precisely in the example, with two of the six deviations being greater than six, and four being less.

The standard error of estimate is a precise measure of just how accurate the derived equation is in estimating actual values. Again, in terms of the current example, one can place a good deal of faith in an equation that is accurate to $6,000 or less two-thirds of the time when dealing with cost figures of this magnitude!

Coefficient of determination: usually symbolized as R^2, or R-squared, is the ratio of variation explained by the equation to total variation in the sample. Total variation is the sum of y^2: squared deviations from the mean of the actual data. This total can be broken down into its two components: that due to the estimated values of the equation, and that representing deviations from these estimated values.

The total variation in your sample is 18478, the sum of squared deviations from estimated values is 143.27. So the ratio of unexplained variation is 143.27/18478, which is equal to .0078. Hence, the ratio of explained variation is 1 minus .0078, which is equal to .9922, and this is the coefficient of determination.

You can arrive at the same answer by taking the deviation of all estimated values from the mean and adding their squares to get the total explained variation, which, in turn, can be divided by the total variation to get the coefficient of determination. But it is easier to arrive at this by the shorter route shown above.

In any event, what the coefficient of determination in this case shows is that 99.22 percent of all the variation in the Y values in this sample can be explained by the derived linear equation. This is a very high degree of association indeed.

You will note that the coefficient of determination is an abstract measure, unlike the standard error of estimate, which is expressed in the original units of Y. Consequently, coefficients of determination can be compared in any test of relationship, irrespective of the units in which the variables are expressed.

Indeed, the two measures supplement each other. The coefficient of determination measures the strength of the relationship, the standard error of estimate how accurate it is in predicting the actual values.

Coefficient of correlation: the square root of the coefficient of determination. In this example, it is the square root of .9922, which is equal to .9961. The usual symbol for this coefficient is R.

It also, of course, is an abstract measure, ranging in value from zero for no correlation to one for perfect correlation. It has been often used in research to compare the relative strength of relationship from one sample to another and for more complex tests of sampling data.

Standard error of b: the standard error of estimate divided by the square root of the variance in X (which has been calculated as the sum of x^2). In the data, this is 5.98 divided by the square root of 806, or 28.39, which is equal to 0.21. Once again, this can be interpreted as a normal deviate, with the odds being two out of three that the true coefficient b in this sample will fall within the range of the derived value of 4.77 plus or minus 0.22.

Thus, the standard error of b gives a direct measure of how much reliance you can place in your derived value, and because an important by-product of the analysis has been to find the marginal cost of production, it is useful to have this gauge of its reliability. In this case, clearly, it is a very accurate measure.

OTHER EQUATIONS

This chapter has dealt only with linear or straight-line relationships between two variables. But sometimes this relationship is nonlinear, and the form of the equation must be changed accordingly. For example, the quadratic equation, $Y = a + bX + cX^2$, will adequately represent many curvilinear relationships between two variables. Here, of course, it is necessary to find the best three coefficients by the least squares method, and the required arithmetic increases accordingly, but the basic concept is the same.

Similarly, it is often possible to change a nonlinear relationship into a linear one by converting the original variables to some other form, such as logarithms. Where appropriate, this is the preferred procedure because one always looks for the simplest form of an equation to describe a given relationship. But again, once the data have been converted to their new form, the calculations are the same as those explained in this chapter.

A good test of the linear nature of a relationship is simply to plot the data with the Y data on the vertical scale and the X data on the horizontal scale. One glance at these plotted points is usually sufficient to judge whether the relationship is a straight or curved line.

But, in general, it is surprising how many relationships in the real world tend to be linear in nature and thus can be fully described by a simple linear equation.

It is also possible, by the method of least squares and a procedure called *multiple regression*, to find the best equation to represent the relationship of a dependent variable to two or more independent variables. While beyond the scope of this book, this is a powerful device for the analysis of complex relationships and enables the researcher to simulate laboratory controls over data in the real world that are essentially uncontrolled.

TECHNICAL NOTE

The arithmetic in the example was simplified by converting the variables to deviations from the mean in each case. In practice, this can be rather onerous when these deviations involve numbers with several decimal figures, but there is an easy way around this.

Using shortcut methods described earlier, you can accomplish this conversion working only with the original numbers. With S representing "Sum of," X and Y in capital letters the original numbers, x and y in small letters their deviations from the mean, and N the number of pairs, you have:

$$S(x^2) = S(X^2) - (SX)(SX)/N$$

$$S(y^2) = S(Y^2) - (SY)(SY)/N$$

$$S(xy) = S(XY) - (SX)(SY)/N$$

This makes the entire calculation much easier to accomplish.

It should also be noted that the coefficients can be derived by solving the following simultaneous equations:

$$S(Y) = a(N) + b(SX)$$

$$S(XY) = a(SX) + b(SX^2)$$

By the same token, a multiple regression problem involving one dependent and two independent variables would require a third term and three simultaneous equations to solve, and so on. A curvilinear relationship of the type $Y = a + bX + cX^2$ is solved the same way with X^2 becoming, in effect, the second dependent variable.

SUMMARY

In this chapter, you have learned to analyze how one variable changes with another and to summarize this relationship in a very simple equation. Further, you have learned how to evaluate the reliability of your results in terms of practical application.

You would be well advised to work out an example of your own to be certain that you have a firm grasp of the arithmetic involved. It is also a good idea to plot the results to see how close the fitted line is to the actual data.

The subject is a big one, and entire books have been written on just one phase of the analytical procedure. But if you have absorbed the basic concept, you are a long way down the road in your ability to understand how many things work in the business world; and this, in turn, can often lead to better control of your working environment.

The example used in this chapter is a good illustration. The analysis has given the plant manager a deeper understanding of the cost structure than could be derived any other way and a practical device for knowing very quickly if costs get out of control at any time.

CHAPTER 19
Trend Analysis

Suppose you are given the task of forecasting sales of your company or division for the next three years, and decide that one rational approach would be to calculate the trend for the past five years and then extrapolate that for three more years. How do you go about this?

The sales record is as follows:

Year	$ 000
1	1,120
2	1,436
3	1,450
4	1,680
5	1,819

Year one represents the first year in the series and Year five the most recent year.

When you plot the data on a chart, they appear to follow a straight-line trend, so your task becomes one of finding the best linear equation to fit the data. For this you turn to the least squares method described in the last chapter. The necessary calculations are shown in Table 19.1, and the data are charted in Figure 19.1.

This analysis tells you that the equation producing the best least squares fit is:

$$\text{EST Y} = 1,008 + 164.2 \, (X)$$

This generates the trend values shown in the table and the chart and also the forecast values shown for the next three years. The analysis also shows you how accurate this equation has been in estimating actual sales in each of the years, in the deviation of actual from estimated values.

The standard error of estimate, computed by the formula set forth in the preceding chapter, is 72 (or $72,000), which means that you

TABLE 19.1
**Fitting a linear equation
By least squares
Y = a + bX**

X	Y	x	y	x^2	y^2	xy
1	1120	−2	−381	4	145161	762
2	1436	−1	− 65	1	4225	65
3	1450	0	− 51	0	2601	0
4	1680	1	179	1	32041	179
5	1819	2	318	4	101124	636
15	7505	0	0	10	285152	1642

$$b = \text{Sum}(xy) / \text{Sum}(x^2) = 164.2$$
$$a = \text{Sum}(Y) - b\text{Sum}(X) / N = 1008.4$$

X	Est. Y	Y	Dev.	Dev.^{2}
1.00	1172.60	1120.00	−52.60	2766.76
2.00	1336.80	1436.00	99.20	9840.63
3.00	1501.00	1450.00	−51.00	2601.00
4.00	1665.20	1680.00	14.80	219.04
5.00	1829.40	1819.00	−10.40	108.16
15.00	7505.00	7505.00	0.00	15535.59

Forecast

6.00	1993.60	
7.00	2157.80	
6.00	2322.00	

can expect the actual total to be within this range of the estimated figure about two-thirds of the time.

While this is not a large amount in relation to the totals involved, you should note that it is fairly large in relation to the average annual increase of 164.2 as shown by the coefficient b.

In brief, if the same underlying trend factors continue, the forecast data may come reasonably close to the mark in terms of total magnitude but may miss the year-to-year changes by a fairly large percentage error.

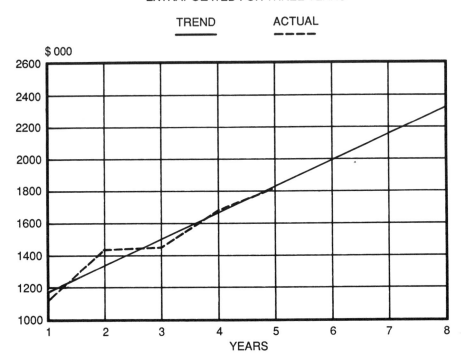

FIGURE 19.1

This is a rather typical situation. Company sales and profits are affected by a wide variety of factors, both external and internal, that vary from one year to another, and it would be rare to find any such data series without erratic fluctuations of some magnitude.

Further, in extrapolating trend lines, one should be aware of the implicit assumption that there will be no change in the basic elements that generated the trend in the first place. Clearly, such changes can occur very rapidly in the business world. Products gain and lose market share. Market demand itself can change. New competitors can spring up, and others can start a price war. And so it goes, on and on.

But all forecasts, of course, are subject to these variables, and the forecast based on extending the trend line may in many cases be more rational than any other approach. In any event, it is often a good starting place: a set of figures that can be modified by users of the data based on their judgment as to what the future may hold.

Apart from this, the trend line has virtue in pinpointing the ups and downs of the data in the past, above and below trend values, for those who wish to study the apparent cause of these fluctuations in prior years as some guide so what may happen in the future.

Under the pressure of current problems, executives tend to neglect this type of study, and this is unfortunate because it can often be very productive in generating ideas and knowledge of considerable practical value. There are lessons in the past, if people will take time to learn them, and failure to do so can sometimes be very costly.

SUMMARY

Once again, the *regression* technique using *least squares* provides a straightforward solution to the problem of finding the best-fitting trend line to a time series, on a completely objective basis.

The simple linear equation works well with data that follow a straight line, but if the trend is in fact curvilinear, then it is necessary to make some changes in the procedure. One such example appears in the chapter that follows.

The arithmetic involved in the regression procedure has been shown here, in all its detail, for the sake of clarity; but in practice, it is easy enough to let machines do this work, either with a simple program on a personal computer or with inexpensive hand-calculators that have been programmed to do the job.

The key fact is that regression technique can be applied in a variety of ways in the business world, and executives who understand it have an extremely valuable tool at their disposal. While the arithmetic may seem tedious, the concept itself is really quite simple, and it can serve you, faithfully and well, throughout your executive career.

CHAPTER 20
Seasonal Variation

Executives live in a dynamic world, and much of the data they work with is of the same nature. Constant change is the norm. Report cards are written every month on these changes in sales, costs, and profit, and management up and down the line is graded accordingly.

But there is a problem in analyzing data of this kind. With few exceptions, all tend to be subject to seasonal variation: fairly regular ups and downs depending on the month and season of the year. These fluctuations are troublesome because they can cloud the underlying trend, which usually is the subject of primary interest.

There is a fairly easy way to isolate and remove this seasonal factor and to help uncover basic trends. You have doubtless seen the effect of this process in something called data "adjusted for seasonal variation," but if you are like most executives, you have only a vague grasp of what is involved in this rather mysterious phrase. This chapter is designed to eliminate the mystery!

The fairly standard practice of dealing with seasonal variation in the business world is to compare data of a certain time period with the same period in a prior year: this month or quarter against the same month or quarter last year, and so on.

This has the virtue of great simplicity. And because it is easy to do and simple to grasp, this practice will certainly continue. Nevertheless, this crude method of comparing data can, and no doubt often does, lead to serious misinterpretation and wrong conclusions.

Take the simple case of a quarterly earnings report, with key figures as follows:

Quarter	Net income in $000		Percent change
	Last year	This year	
1	$2,104	$2,378	13.0
2	3,334	3,184	− 4.5

Surely, nothing can be simpler than this. The company had a good first quarter and a bad second quarter. The earnings trend started out positive and ended up negative. That is the obvious interpretation. But is it correct?

The answer is that it may or may not be correct. From the data as given, there really is no way to tell. And there is a simple reason for this. If you think about it, you will see that the obvious interpretation assumes that *last year was normal*. This may or may not be the case.

In other words, the first quarter may be up due wholly to the fact that the same quarter last year, for some reason, was unusually low. And the second quarter this year may be down because the same period last year was unusually high. Or there may have been some combination of these things. On all this, the data as given are essentially silent.

For example, suppose that you know what the first and second quarter earnings should have been last year based on their normal percentage of the year's total, which gives the following comparison:

Quarter	Net Income in $000		
	Normal last year	Actual this year	Percent change
1	$2,198	$2,378	8.2
2	3,027	3,184	5.2

This picture is altogether different from the one based on the original numbers. There is no basis here for any inference that earnings were in a strong positive trend in the first quarter and then deteriorated to a negative trend in the second quarter, as the original data would suggest.

Managers will readily concede that all this is possible. They know that the prior year may be abnormal, but, with no way to quantify this, their tendency is simply to ignore it. The effect is to attribute all changes to the current period and to implicitly assume normality in the prior period.

I repeat, these comparisons with the same period last year are commonplace and will continue, but you should always be aware of the pitfalls in interpreting such data.

ADJUSTMENT FOR SEASONAL VARIATION

When you are dealing with an important time series where thorough analysis is justified, the techniques described in this chapter are a definite step in the right direction. Adjustment for seasonal variation may not give you the final answer in analyzing data of this kind, but it can often be most helpful in clarifying trends and arriving at a proper interpretation.

The typical method used to adjust for seasonal variation is based on a very simple concept. A fair amount of arithmetic is involved, but the process itself is quite logical and straightforward. Usually you are dealing with monthly data, and what you want to find is the normal ratio of a given month to the average month of the year. Here is the procedure:

1. For whatever time series is involved, monthly data are collected for a number of years—preferably ten or more—in the past.
2. Each month is expressed as a percentage of the moving average based on the 12-month total centered on that month.
3. An average is taken of these percentages to find the average percent of each month to the centered moving average. An adjustment is then made pro rata, if necessary, so that these averages will add up to 1200 for the average year they represent.

About the only complication in all of this is in the centering of the moving total and moving average, because the calendar year is not well adapted for this purpose. If there were a 13-month year, it would be centered on the seventh month and there would be no problem. But life is not so simple and a 12-month year is centered at the end of the sixth month. The total from January to December, for example, is centered at the end of June. All this is a bit of a nuisance, but it can be overcome rather easily.

The sales of a company, as shown in Table 20.1, will illustrate the entire procedure over a ten-year period.

The first step is to compute moving 12-month totals, which appear in Table 20.2. These are centered at the end of the month they are listed under, which is the sixth month in the 12-month period in each case.

In other words, the total from July of the prior year to June of the current year is listed under December of the prior year, and it is

TABLE 20.1
Sales in $000

Year	Jan	Feb	Mar	Apr	May	Jun	Jul	Aug	Sep	Oct	Nov	Dec
1977	0	0	0	0	0	0	872	1024	1441	1616	1343	1122
1978	1086	1120	1234	1476	1676	1065	1077	1159	1675	1896	1526	1354
1979	1276	1323	1503	1672	1907	1218	1163	1355	1852	2086	1685	1559
1980	1436	1490	1707	1899	2142	1399	1305	1497	2022	2332	1949	1766
1981	1591	1679	1931	2095	2493	1597	1508	1646	2286	2593	2176	1880
1982	1759	1868	2109	2361	2763	1722	1621	1820	2479	2855	2373	2062
1983	1942	2002	2302	2600	2920	1870	1733	1957	2767	3114	2617	2288
1984	2136	2224	2492	2747	3213	2110	1937	2142	2933	3406	2790	2496
1985	2271	2428	2711	3034	3523	2282	2043	2286	3143	3600	2969	2658
1986	2445	2535	2883	3250	3811	2415	2167	2474	3400	3870	3218	2876
1987	2641	2719	3104	3522	4063	2574	0	0	0	0	0	0

TABLE 20.2
12-month moving totals
Centered at the end of each month*

Year	Jan	Feb	Mar	Apr	May	Jun	Jul	Aug	Sep	Oct	Nov	Dec
1978	0	0	0	0	0	0	0	0	0	0	0	15075
1979	15280	15415	15649	15929	16112	16344	16534	16737	17006	17202	17433	17586
1980	17672	17868	18045	18235	18394	18599	18759	18926	19130	19357	19592	19773
1981	19915	20057	20227	20473	20737	20944	21099	21288	21512	21708	22059	22257
1982	22460	22609	22873	23134	23361	23475	23643	23832	24010	24276	24546	24671
1983	24784	24958	25151	25413	25610	25792	25975	26109	26302	26541	26698	26846
1984	26958	27095	27383	27642	27886	28112	28306	28528	28718	28865	29158	29398
1985	29602	29787	29953	30245	30418	30626	30761	30965	31184	31471	31781	31953
1986	32059	32203	32413	32607	32786	32948	33122	33229	33401	33617	33905	34038

*The 15075 sum listed for Dec 1978 is the 12-month total ending Jun 1978.
The 15280 sum listed for Jan 1979 is the 12-month total ending Jul 1978.
And so on.

centered at the end of December, the sixth month in this 12-month period.

Similarly, the 12-month total listed under January includes the period from August of the prior year to July of the current year and is centered at the end of January.

Now, to get the moving total centered in the middle of January, simply average the total centered at the end of December and the total centered at the end of January. The result of this process is shown in Table 20.3.

All this sounds more complicated than it really is. This extra effort to center the data by averaging the adjacent moving totals is not likely to have much effect on final results, and some analysts may choose to eliminate this step accordingly. But the little extra effort is not much trouble, especially if computers are used to do the arithmetic, and probably most analysts would include it accordingly.

The next step, in Table 20.4, is to divide the moving total by 12 to get the centered moving average, and then, as in Table 20.5, to calculate for each month its percent of this average. When you add up these percentages and divide by the number of years involved, you get the desired percentage of each month to the average month of the year.

One final step is to adjust these percentages to add up to 1200. In this example they come very close, to 1199.5. The adjustment process is simply to multiply each percentage by the ratio of 1200/Sum, which in this example is 1200/1199.5. This is a negligible adjustment in this case, as can be seen in the minor changes reflected in the final averages shown in the lower half of the table.

SEASONAL INDEXES

You now have the results you have been looking for: seasonal indexes with each month expressed as a percentage of the average month of the year. You see, for example, that January tends to be below norm, with a seasonal index of 86.2 percent, while May is well above norm, with a seasonal index of 127.5 percent, and so on. When plotted in Figure 20.1, these indexes paint a very clear picture of the seasonal fluctuation in the sales data.

As the next step in your analysis, using these seasonal indexes you can now go back and adjust all of the monthly totals for seasonal variation. All you need do is divide the original data by the

TABLE 20.3
12-month moving totals
Centered on each month*

Year	Jan	Feb	Mar	Apr	May	Jun	Jul	Aug	Sep	Oct	Nov	Dec
1979	15178	15348	15532	15789	16021	16228	16439	16636	16872	17104	17318	17510
1980	17629	17770	17957	18140	18315	18497	18679	18843	19028	19244	19475	19683
1981	19844	19986	20142	20350	20605	20841	21022	21194	21400	21610	21884	22158
1982	22359	22535	22741	23004	23248	23418	23559	23738	23921	24143	24411	24609
1983	24728	24871	25055	25282	25512	25701	25884	26042	26206	26422	26620	26772
1984	26902	27027	27239	27513	27764	27999	28209	28417	28623	28792	29012	29278
1985	29500	29695	29870	30099	30332	30522	30694	30863	30175	31328	31626	31867
1986	32006	32131	32308	32510	32697	32867	33035	33176	33315	33509	33761	33972

*The 15178 sum listed for Jan 1979 is the average of the two 12-month totals ending Jun 1979 and Jul 1979. The 15348 sum listed for Feb 1979 is the average of the two 12-month totals ending Jul 1979 and Aug 1979. And so on.

TABLE 20.4
Moving average

Year	Jan	Feb	Mar	Apr	May	Jun	Jul	Aug	Sep	Oct	Nov	Dec
1978	1265	1279	1294	1316	1335	1352	1370	1386	1406	1425	1443	1459
1979	1469	1481	1496	1512	1526	1541	1557	1570	1586	1604	1623	1640
1980	1654	1666	1679	1696	1717	1737	1752	1766	1783	1801	1824	1847
1981	1863	1878	1895	1917	1937	1952	1963	1978	1993	2012	2034	2051
1982	2061	2073	2088	2107	2126	2142	2157	2170	2184	2202	2218	2231
1983	2242	2252	2270	2293	2314	2333	2351	2368	2385	2399	2418	2440
1984	2458	2475	2489	2508	2528	2544	2558	2572	2590	2611	2636	2656
1985	2667	2678	2692	2709	2725	2739	2753	2765	2776	2792	2813	2831
1986	2842	2855	2873	2895	2917	2936	2954	2969	2986	3007	3029	3046

TABLE 20.5
Percent of moving average
Centered on each month*

Year	Jan	Feb	Mar	Apr	May	Jun	Jul	Aug	Sep	Oct	Nov	Dec
1978	85.9	87.6	95.3	112.2	125.5	78.8	78.6	83.6	119.1	133.0	105.7	92.8
1979	86.9	89.3	100.4	110.6	125.0	79.0	74.7	86.3	116.8	130.1	103.8	95.0
1980	86.8	89.5	101.7	112.0	124.7	80.6	74.5	84.8	113.4	129.5	106.9	95.6
1981	85.4	89.4	101.9	109.3	128.7	81.8	76.8	83.2	114.7	128.9	107.0	91.7
1982	85.4	90.1	101.0	112.1	130.0	80.4	75.2	83.9	113.5	129.7	107.0	92.4
1983	86.6	88.9	101.4	113.4	126.2	80.1	73.7	82.6	116.0	129.8	108.2	93.8
1984	86.9	89.9	100.1	109.5	127.1	83.0	75.7	83.3	113.3	130.5	105.9	94.0
1985	85.1	90.7	100.7	112.0	129.3	83.3	74.2	82.7	113.2	128.9	105.5	93.9
1986	86.0	88.8	100.3	112.3	130.7	82.2	73.4	83.3	113.9	128.7	106.3	94.4
Avg.	86.1	89.4	100.3	111.5	127.5	81.0	75.2	83.7	114.9	129.9	106.3	93.7

Adjusted seasonal indexes

Month	Pct. of year	Pct. of average
Jan	7.18	86.2
Feb	7.45	89.4
Mar	8.36	100.4
Apr	9.29	111.5
May	10.63	127.5
Jun	6.76	81.1
Jul	6.27	75.2
Aug	6.98	83.8
Sep	9.58	114.9
Oct	10.83	130.0
Nov	8.86	106.3
Dec	7.82	93.8
Sum	100.00	1200.0

150 BASIC BUSINESS STATISTICS FOR MANAGERS

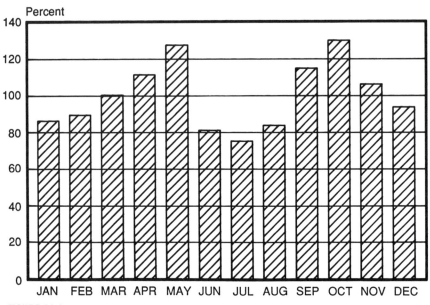

FIGURE 20.1

appropriate seasonal index in ratio form: January divided by .862, February by .894, and so on. Results are shown in Table 20.6 and are graphically portrayed in Figure 20.2.

One glance at the chart is sufficient to show that the analysis and adjustment for seasonal variation has been highly effective in this case. Once the rather violent seasonal fluctuation is removed, the underlying trend becomes very clear indeed.

NO RESTRICTION IN DATA COMPARISON

Now, here is the significant thing about data adjusted for seasonal variation. You are no longer restricted to comparing data for one time period with the same time period in prior years. Any month can be compared with any other month. And this broadens your analytical horizon enormously.

TABLE 20.6
Sales adjusted for seasonal variation ($000)

Year	Jan	Feb	Mar	Apr	May	Jun	Jul	Aug	Sep	Oct	Nov	Dec
1978	1260	1253	1229	1324	1315	1313	1432	1383	1458	1458	1436	1443
1979	1480	1480	1497	1500	1496	1502	1547	1617	1612	1605	1585	1662
1980	1666	1667	1700	1703	1680	1725	1735	1786	1760	1794	1833	1883
1981	1846	1878	1923	1879	1955	1969	2005	1964	1990	1995	2047	2004
1982	2041	2089	2101	2117	2167	2123	2156	2172	2158	2196	2232	2198
1983	2253	2239	2293	2332	2290	2306	2305	2335	2408	2395	2462	2439
1984	2478	2488	2482	2464	2520	2602	2576	2556	2553	2620	2625	2661
1985	2635	2716	2700	2721	2763	2814	2717	2728	2735	2769	2793	2834
1986	2836	2836	2872	2915	2989	2978	2882	2952	2959	2977	3027	3066

152 BASIC BUSINESS STATISTICS FOR MANAGERS

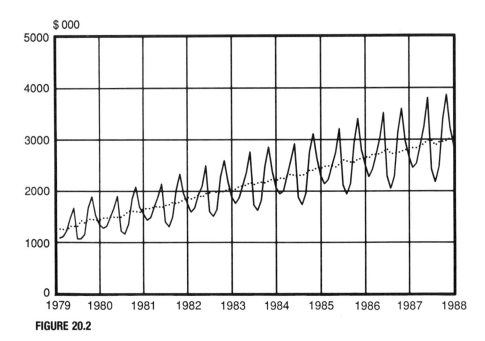

FIGURE 20.2

For example, take the last six months of 1986 and look at the actual versus adjusted totals for each month:

	Sales in $000	
1986	Actual	Adjusted
Jul	2,167	2,882
Aug	2,474	2,952
Sep	3,400	2,959
Oct	3,870	2,977
Nov	3,218	3,027
Dec	2,876	3,066

Now, what kind of sense can you make from the actual figures in terms of sales trend? The answer, of course, is really no sense at all. By contrast, look at the adjusted figures, and you get a very clear

TABLE 20.7
Sales adjusted for seasonal variation
1978 = 100

Year	Jan	Feb	Mar	Apr	May	Jun	Jul	Aug	Sep	Oct	Nov	Dec
1978	93	92	90	97	97	96	105	102	107	107	105	106
1979	109	109	110	110	110	110	114	119	118	118	116	122
1980	122	122	125	125	123	127	127	131	129	132	135	138
1981	136	138	141	138	144	145	147	144	146	146	150	147
1982	150	153	154	155	159	156	158	159	158	161	164	161
1983	165	164	168	171	168	169	169	171	177	176	181	179
1984	182	183	182	181	185	191	189	188	187	192	193	195
1985	193	199	198	200	203	207	199	200	201	203	205	208
1986	208	208	211	214	219	219	212	217	217	219	222	225

picture of what has been happening to sales in this six-month period. This simple example clearly illustrates the great power of seasonal adjustment.

Going back to the adjusted data in Table 20.6, you may wish to make these figures even easier to read by putting them in index form. This is done in Table 20.7 by choosing 1978 as the base and showing all the data as a percentage of the 1,362 monthly average in the base year. Of course, you may select any other year, or average for several years, as the base for this purpose.

SPECIAL TECHNIQUES

Seasonal indexes for other time periods, such as weeks or quarters, can be developed by the same techniques used in the example for monthly data.

These techniques are usually adequate for most situations, but some refinement of the original data may sometimes be in order. A good illustration is advertising linage in a newspaper, which fluctuates not only by season of the year but by day or week. Linage tends to be heavy on Thursday and Sunday, relatively light on Monday and Saturday. Thus it makes a good deal of difference, for example, whether a month has four or five Sundays; and this variation constantly occurs.

One solution to this is to get the average linage for each day of the week in a given month and then add these averages per day to get a weekly total to represent that particular month. The seasonal adjustment is then worked out using these weekly totals for each month as the basic data involved. Seasonal indexes thus derived then correct not only for monthly fluctuation but also for variation in days of the week in any month.

It sometimes happens that seasonal variation shifts over time, and it is possible to measure and adjust for this; but this requires a bit more skill than the average executive is likely to bring to the task and is more appropriate for a professional analyst.

SUMMARY

Time series give a moving picture of the dynamic business world, but that picture is often blurred by seasonal fluctuation of one kind or another. Where it is important to measure trends on a current basis and avoid incorrect interpretations, the adjustment of data for seasonal variation can be exceedingly helpful.

The method of deriving seasonal indexes is quite simple and straightforward, and the only real difficulty is in the amount of arithmetic involved. But it is really very easy to program a computer to do the necessary arithmetic. Once the program is written, the only real work left is the purely clerical one of developing and entering the basic data for prior years.

This ability to develop seasonal indexes and to adjust data for seasonal variation is a valuable tool, and the effort required is relatively minor in terms of the benefit that can be derived.

CHAPTER 21
Compound Growth Rates

In the business world, you often read about something growing at some compound growth rate over a specified period of time. In the stockholders' letter in an annual report, for example, you are likely to see some statement about the company's earnings per share growing at a compound growth rate of xx percent over the past xx years.

Clearly, this is a very useful summary device. It is handy indeed to be able to characterize the growth in a given time series by a single number for comparison with similar numbers calculated for other series. It is helpful to stockholders, for example, to know how the growth of their company compares with that of other companies.

Now, the odd thing is that no matter how familiar this descriptive measure of a time series is, I doubt very much that most executives have any real knowledge of how it is computed. They know pretty much what it means from the name itself, but if they tried to calculate such a measure, the chances are that they would end up with the wrong answer.

For example, from the data on the sales of your company for the past six years shown below, you are asked to calculate the annual compound growth rate. The years are coded here, so that Year one is the first year in the series and Year six is the last year:

Year	$ Millions
1	80
2	120
3	136
4	127
5	130
6	175

You might decide to answer the question by this reasoning: since the company started with sales of $80 million and ended up with sales of $175 million over a five-year period, why not find the compound rate that will make 80 grow to 175 in five years. This is certainly

156

one way to calculate growth rates, and it is easy enough to do on your little hand calculator.

When 175 is divided by 80 you get a ratio of 2.1875, and when you take the one-fifth root of this, you get 1.1695. This is the ratio by which you must multiply each preceding number, beginning with 80, to arrive at 175 at the end of five years. By subtracting one from this ratio and putting the remainder in percentage form, you translate this to an annual compound growth rate of 16.95 percent. The arithmetic works out fine, but is it the right solution?

The answer is that this solution can be quite misleading, but I have a suspicion, nevertheless, that many of the compound growth rates you read about are calculated this way by people who don't understand the basic problem involved in this rather simple and straightforward approach.

The problem is simply this. What you are looking for is a figure that will summarize the growth of the entire series. When you confine your calculation to just the first and the last number, you are ignoring everything in between, and that is a very poor averaging system indeed.

To illustrate this, look at the two series of numbers below:

Year	A	B
1	1,000	1,000
2	− 2,000	1,200
3	− 3,000	1,400
4	1,600	1,600

These two series start and end the same way, but would you conclude that they have the same compound growth rate? If the two sets of numbers reflect the net income or loss of two corporations, would you want to characterize their relative growth rates as being the same over this period and be equally willing to invest in the stock of one versus the other?

The answers are pretty evident from the table itself, and even more dramatically clear when the two series are charted in Figure 21.1. Quite obviously, trying to summarize the growth in a given time series by a calculation based on nothing more than the ratio of the last to the first figure can lead to logical nonsense of the first order.

BASIC BUSINESS STATISTICS FOR MANAGERS

TWO SERIES WITH THE SAME COMPOUND GROWTH RATE
ACCORDING TO AN ERRONEOUS SHORTCUT FORMULA

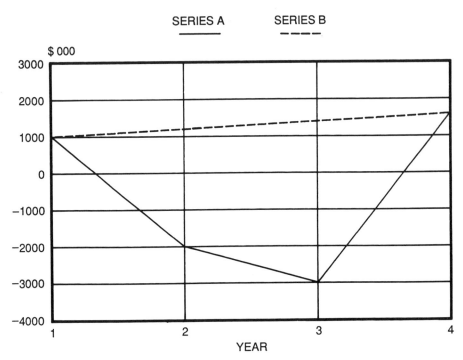

FIGURE 21.1

Now, if that is so, the question is how to overcome this problem and somehow take account of every figure in the series while giving equal weight to each. For the solution to this, you need to turn to regression technique, with one modification in the procedure described in the earlier chapter on Regression.

Again, you need to fit a linear equation to a series using the least squares method, but in this case you need to convert the series into logarithms. The reason for this is that adding logarithms is the same as multiplying the original numbers. The linear equation gives the best value of the coefficient b, which is the amount added to get the line of best fit to the data.

For example, if you multiply each preceding number by the ratio of 1.10, you are in effect compounding at the rate of 10 percent per year. The natural log of 1.10 is .0953. Thus, if you start with the log of an original number and add .0953 per year, the antilog of the number you end up with is the original number compounded at 10 percent a year over the given period.

Try this with a few numbers to see how it works:

Year	Number	Log	Increase
1	1,000	6.9078	–
2	1,100	7.0031	.0953
3	1,210	7.0984	.0953

The increase of .0953 in the log each year is the same as multiplying the preceding original number by 1.10, which is the same as adding 10 percent to it each year.

So when you fit the linear equation to the logs of the series data and then find the coefficient b, you discover the multiplying ratio to apply to each preceding number in the way that will best fit all the data. The usual regression calculations are shown in Table 21.1, and are the same as before except that the actual Y values have been converted into natural logarithms.

You see from this that the coefficient b is .1167299, with an antilog of 1.123816. The latter is the ratio of each value in the fitted line to the previous value and represents, of course, a growth rate of 12.38 percent. This is significantly different from the rate of 16.95 percent produced by the shortcut and erroneous calculation.

In the lower half of Table 21.1 are the estimated values from the equation versus actual values in both logs and antilogs, and the later are plotted in Figure 21.2, which is a graphical representation of the entire procedure.

SUMMARY

There really is only one proper way to calculate the compound annual growth rate of a time series while giving equal weight to each item in the series, and that is the procedure described above. The shortcut approach based on only the first and last item in the series can be quite misleading and should not be used.

Again, the arithmetic here appears more formidable than it is in practice. Small hand calculators will now convert numbers into logarithms or convert logs back into their antilog values at the punch

TABLE 21.1
Fitting a linear equation
By least squares
Y = a + bX

X	Y	x	y	x^2	y^2	xy
1.0000	4.3820	−2.5000	−0.4444	6.2500	0.1975	1.1111
2.0000	4.7875	−1.5000	−0.0389	2.2500	0.0015	0.0584
3.0000	4.9127	−0.5000	0.0862	0.2500	0.0074	−0.0431
4.0000	4.8442	0.5000	0.0178	0.2500	0.0003	0.0089
5.0000	4.8675	1.5000	0.0411	2.2500	0.0017	0.0616
6.0000	5.1648	2.5000	0.3384	6.2500	0.1145	0.8459
21.0000	28.9587	0.0000	0.0000	17.5000	0.3230	−2.0428

$$b = \text{Sum}(xy) / \text{Sum}(x^2) = .1167299$$
$$a = \text{Sum}(Y) - b\text{Sum}(X) / N = 4.417888$$
$$\text{Growth Rate} = \text{antilog of } b = 1.123816$$

	Estimated		Actual		Deviation		
X	Log Y	Y	Log Y	Y	Log Y	Y	Dev^2
1.00	4.53	93.19	4.38	80.00	−0.15	13.19	0.02
2.00	4.65	104.73	4.79	120.00	0.14	−15.27	0.02
3.00	4.77	117.69	4.91	136.00	0.14	−18.31	0.02
4.00	4.88	132.27	4.84	127.00	−0.04	5.26	0.00
5.00	5.00	148.64	4.87	130.00	−0.13	18.65	0.00
6.00	5.12	167.05	5.16	175.00	0.05	−7.96	0.00
21.00	28.96		28.96		0.00		0.08

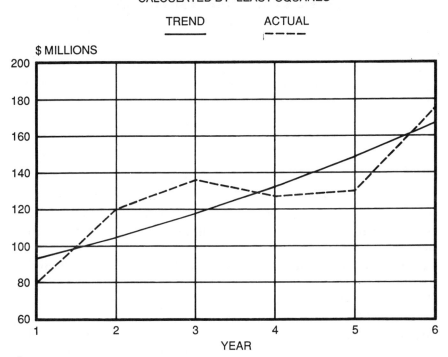

FIGURE 21.2

of a button. And the entire least squares procedure is programmed in some of these little machines for easy operation.

But whether you want to do any actual calculating or not, it is important for you to understand the concept and to be able to monitor others who may do the work for you. This is the only way to be quite certain that they are following the right procedure, and failure to do so, as you have seen, can lead to some very gross errors indeed.

CHAPTER 22
Index Numbers

Executives need to have some grasp of index numbers, how they are created and what they mean. They can be used for many purposes: to clarify relationships between numbers and to measure highly complex phenomena ranging from stock prices to the gross national product.

It is a simple task to create index numbers. You can easily learn to do it, and this ability can be very helpful if you have anything to do with preparing business reports.

Converting original data to index numbers is especially useful in comparing changes over time in two or more series. For example, consider the following sales data for three company divisions:

Year	Sales in $000		
	Div. A	Div. B	Div. C
0	1,328	831	3,982
1	1,511	917	4,776
2	1,643	1,213	4,325
3	1,519	1,086	5,687
4	1,934	1,624	6,001
5	1,822	1,761	6,225

All the numbers are here, but it is quite difficult to get any clear perception of the relative growth of each division compared with the other two, year by year, throughout this period.

Now, express the sales of each division each year as a percent of its sales in the base year 0:

Sales: Year 0 = 100

Year	Div. A	Div. B	Div. C
0	100	100	100
1	114	110	120
2	124	146	109
3	114	131	143
4	146	195	151
5	137	212	156

And I think you will agree that these new numbers give a much clearer picture of what has happened to the annual sales of each division over this period and make it fairly easy to compare one against another.

But you should be aware of some pitfalls and potentials for deception in this procedure, mainly having to do with the base period selected. If, for example, sales in one of the divisions were unusually low in the base period, sales in succeeding years will look quite high relative to that base.

Perhaps the best way to avoid this is to select several years, perhaps three, and make their average the base for the relative numbers. It may also be useful, if the data are charted, to use semilog charts where the slope from one point to another is proportionate to the rate of change, irrespective of the base period.

Typically, of course, index numbers are used to summarize far more complex data than those in this example, such as national indexes of wholesale prices, industrial production, and consumer prices, and the procedures used become equally complex.

However, a very simplified example of a company with two products, one measured in tons and the other in individual units, can illustrate the general concept:

Sales and Production Record in 000

Year	$ Sales	Product A (in tons)	Product B (in units)
1	12,860	7,500	2,680
2	15,492	8,210	2,590
3	15,967	6,420	3,270
4	22,267	8,730	4,160
5	25,083	9,150	4,650

164 BASIC BUSINESS STATISTICS FOR MANAGERS

Now, to get a combined measure of production you clearly cannot add tons and individual units. First, you need to convert them into some common measure, and one simple way to do this is to use the price per ton and the price per unit in the initial year. If the price per ton was $100 and the price per unit was $200 in the first year and you multiply all the production units by these respective prices each year throughout the period, you get a constant dollar value of production as shown below:

Year	Constant dollar value of production	Sales	Price index
1	$12,860	$12,860	100.0
2	13,390	15,492	115.7
3	12,960	15,967	123.2
4	17,050	22,267	130.6
5	18,450	25,083	136.0

By valuing each element of production at its base period price, you in effect have converted production to a constant dollar value, and when you divide actual sales by this measure of unit production, you arrive at the average price index shown in the last column. When all of this is put in index form, with the base year = 100 in each case, you get the following summary table:

Year	Indexes		
	Sales	Production	Price
1	100.0	100.0	100.0
2	120.5	104.1	115.7
3	124.2	100.8	123.3
4	173.1	132.6	130.6
5	195.0	143.5	136.1

And that, in the current vernacular, is a pretty neat solution. At any point one can see at a glance the relative contribution of production versus price to the increase in dollar sales volume from the base period.

But the key point is that when you want to combine series with dissimilar units, you must use some weighting system. In this example, unit prices in the base period were used to weight the production units, but unit prices in some other period, such as the current year, could have been chosen instead. You should be aware of the fact that the resulting measure of production will change with each such change in weights.

You may ask: how can this be? Should not the same numbers always produce the same answer? But when you change weights and assign different values to the number, you are in effect asking a different question each time, and it is to be expected that the answer in the form of combined totals will vary accordingly.

Unit prices are used to weight quantities in deriving a production index, but the opposite applies in creating a price index, where quantity is used to weight unit prices. Perhaps the prime example—and one of great importance—is the national Consumer Price Index.

CONSUMER PRICE INDEX

No one really knows how many billions of dollars are riding on every point change in the Consumer Price Index, which is published monthly by the Bureau of Labor Statistics of the U.S. Department of Labor.

Many years ago, I had the privilege of serving on an Advisory Committee to the Bureau, and the opportunity to see at firsthand some of the basic problems that are inherent in any such index.

Those responsible for this index, knowing full well the multibillion dollar impact of their product, have no easy task to perform, and business executives in particular would do well to know more about the subject.

The Consumer Price Index (CPI) is the one normally used for cost-of-living adjustments (COLA clauses) in labor contracts, social security benefits, and many other entitlement programs.

It is worth noting that the Bureau of Labor Statistics does not label its product a Cost-of-Living Index, but rather a Consumer Price Index, and that difference is a good deal more than mere semantics even though, in practice, it is almost totally ignored.

Put more bluntly, all these COLA adjustments in labor contracts and government benefit programs are designed to offset, in whole or in part, changes in cost of living due to inflation. But the index used for that purpose is designed to measure something else entirely—i.e., changes in certain consumer prices.

It can be argued, of course, that these two things are highly correlated, and clearly there is some merit in that. But perhaps the most cogent argument—which you are not likely to hear—is that there is no realistic way to measure changes in cost of living, and the only thing left is to come up with some measure of changes in the consumer price level as a proxy device.

The formal description of the Consumer Price Index by the U.S. Bureau of Labor Statistics is as follows:

> The CPI is a measure of price change for a fixed market basket of goods and services of constant quantity and quality purchased for consumption. It is essential to update that market basket periodically so that the CPI reflects price changes of items currently purchased by consumers.

The basic technique is to conduct an expenditure study in urban areas about once every ten years to derive a "market basket" of items purchased. Then, in subsequent surveys the quantity of each such item in the market basket is multiplied by its current price. The index then reflects the dollar total so obtained for the overall market basket.

This description is sufficient in itself to indicate the magnitude and complexity of this undertaking. In addition to the many sampling problems in this procedure, there are a host of other difficulties involved.

There is the question, for example, of substitution. In real life, families and individuals tend to react to price changes by substituting one item for another. If the price of beef goes up, they may buy chicken instead, and so on. There is no practical way to deal with this in a procedure where weights are based on a fixed market basket of items.

Also quite troublesome is how to equate homeowners and renters in the market basket and compute some kind of rent equivalent for the former.

Quality changes are also very difficult to evaluate. How is it possible to compare the price of an automobile today with the price ten or twenty years ago? Clearly a price per pound or per unit of horsepower makes no sense in this context. And, of course, there are many new products—such as home computers, VCR's, and the like—that were not available at all in earlier periods.

The Bureau of Labor Statistics is quite candid in discussing this basic problem:

> One of the most difficult problems for those who compile price indexes is that of quality change. Products and services change constantly, and new items replace old ones on the market. There is a large body of literature on the effect of quality change on Consumer Price Indexes. Most of these studies show mixed results. Although it is generally agreed that quality adjustment error exists, the extent of the error, and, indeed, even its direction, are not known.

All of this should be quite sufficient to indicate that it is hard enough to measure changes in consumer prices, much less deal with changes in cost-of-living, whatever that term may mean.

And all of this doesn't even take into account the great ambiguity that a person's standard of living tends to rise with income and the cost of maintaining that standard goes up accordingly, and all of us tend to find it is very difficult to distinguish between the two.

In any event, the Consumer Price Index is the only real measure of living costs and about the only device available for adjusting dollar incomes over time to dollars of equal purchasing power. As this example illustrates, this deflation process is very simple:

Year	CPI	Salary	Constant dollars
0	100.0	$50,000	$50,000
1	107.8	55,000	51,020
2	114.3	60,000	52,493

The procedure here is to recalculate the CPI by dividing the figure for each year by that for the base year in order to make the base year 100. The income for each year is then divided by this index in ratio form to get the adjusted income in base year constant dollars. For all the reasons that have been mentioned, this may be a crude adjustment, but it does indicate, in this case, that the specified salary has done little more than keep up with inflation, with only a small gain in real purchasing power.

If your own salary is involved, only you can decide whether it is judicious to show this analysis to your supervisor, but it may at least have some propaganda value on the home front in the family battle of the budget!

In any event, this is the deflation process using any price index. It is not required that the price index be restated with a new base year as 100, but this is often convenient and may make the results easier to understand.

SUMMARY

Index numbers can often be used to measure and clarify important concepts in the business world, and all executives should have some grasp of the subject.

While there is far more to index number construction than has been covered in this brief space, the fundamentals outlined here should help you understand the key principles involved.

CHAPTER 23
Significant Figures

Most executives at one time or another need to prepare written reports for others, and these must often include numerical data. Those who read such reports can be much influenced, not only by the content, but also by the style of presentation. Therefore, it is important for executives to know a few guidelines on how numbers should be displayed for others to read and understand.

When I was a young man, just beginning to work on economic and statistical research, my only mechanical aid was a slide rule and I performed some fairly difficult calculations with this rather ingenious but primitive device. It could be done, but it was highly tedious and conducive to a good deal of eyestrain.

And then I fell heir to a second-hand calculator, a complete antique by today's standards, but it could multiply and divide and I was thrilled to see it grind out these numbers, slowly but surely, digit by digit. Compared to the slide rule, it was pure magic, and I remember thinking that no one could possibly improve on this great invention!

So thrilled was I at the marvels of this new mechanical aid that I began to produce reports with percentages carried to four decimal points, and took great pride in this impressive product. Somewhat later, when I was both older and wiser, I was quite embarrassed by all this, and hoped that all these amateurish reports had long since been discarded and would never reappear.

The moral of this little story is that the proper presentation of numbers, in terms of how many digits are displayed, is one way that an amateur can be spotted, and thus it is worth giving some thought to the subject. Let us take this number as an example:

$$\$173{,}224{,}457.29$$

The key question is how many digits in this number really mean anything in terms of practical significance. (Disregard accounting data, where this kind of precision may sometimes be required.) The answer is that most of the digits are quite meaningless in a practical sense.

When dealing in hundreds of millions of dollars, the number of pennies or dollars or even thousands of dollars must be considered as quite immaterial. Very little in terms of real meaning would be lost if this number were condensed to:

$173 million

These three digits are really quite sufficient, and all the other digits simply confuse the issue.

The need to cut down on the number of digits to be reported becomes even more imperative when there are a series of such numbers, as shown below:

Division	Year 1	Year 2	Change
A	3,428,176	3,909,671	481,495
B	2,678,301	2,491,223	−187,078
C	4,719,866	4,211,307	−508,559
D	2,399,607	2,741,212	341,605
E	5,042,319	4,881,637	−160,682
F	3,882,171	4,235,192	353,021

When the meaningless digits are dropped, the gain in clarity is quite dramatic:

	(In 000)		
Division	Year 1	Year 2	Change
A	3,428	3,910	482
B	2,678	2,491	−187
C	4,720	4,211	−509
D	2,400	2,741	341
E	5,042	4,882	−160
F	3,882	4,235	353

While there is no absolute rule that governs the truncation of digits in a series of numbers, a pretty good rule of thumb is to retain at least three or four significant digits in the smallest number involved. If this rule is followed, any error potential will be negligible.

Significant Figures   171

Now, a word about rounding numbers. The normal convention is to round down the number 12.449 to 12.4 but to round up the number 12.450 to 12.5, so that the maximum error is one-half a digit in the rounded figure.

This works quite well unless you have a series of percentages that should add to 100 percent and the normal rule leaves you with a series of numbers that do not add up to this total. In this case, you may elect to adjust the rounded numbers so they will add to the proper total.

Here is an example:

Number	Percent	Rounded
1	24.449	24.4
2	61.324	61.3
3	14.227	14.2
6	100.000	99.9

In an extensive report with many percentage distributions, the report writer may simply leave these figures alone and explain with a footnote that percentages may not add to 100 due to rounding.

But in most business reports it is better to avoid this by adjusting the rounded numbers, and essentially there are two ways to do this:

1. Adjust the number that comes closest to the normal rounding rule. In this example, 24.449 would be rounded to 24.5. Or,
2. Adjust the largest number in the group. In this example, 61.324 would be rounded to 61.4.

The first method may look more logical than the second, but in terms of relative error the second may in fact be superior and, indeed, is so in this case.

Rounding up the smaller number represents an error of .051 on a total of 24.449, which is relatively more than the error of .076 on a total of 61.324 produced in rounding up the larger number.

But, from a practical viewpoint, the error is negligible in either case, and it is a matter of indifference which method is chosen.

There is one caution to be observed in rounding numbers. If they are to be used to multiply or divide each other, they should have at least one more significant figure than desired in the final result. Here is an example:

$$1.2 \times 1.2 = 1.44$$

$$1.249 \times 1.249 = 1.56$$

$$1.15 \times 1.15 = 1.32$$

The numbers on the first line are the rounded equivalent of those in the next two lines, but the extra digits are needed to get the true product when the numbers are multiplied.

SUMMARY

The key lesson of this chapter is that since most business reports are designed to be read by executives under heavy time pressure, clarity of presentation is of the essence.

One way to improve that clarity is to use the simple rules that have been outlined here for dropping insignificant digits from numbers being presented. Based on long experience in both preparing and reading such reports, I can assure you that the final result is well worth the small effort involved.

CHAPTER 24
Statistics and Scientific Method

Executives who become skilled in dealing with numbers will use them for a variety of purposes. Much of this work will be purely descriptive in nature: such things as how several divisions compare in sales growth and profitability, the rate of return on a potential capital outlay, and so on.

The other major category of use is in statistical inference, where certain populations are sampled and the sample data are used to estimate various aspects of the population as a whole. Marketing surveys are typical examples of this application.

Beyond this, statistics also enter into a broader system of inquiry under the heading of *scientific method*, where the attempt is to find relationships of some generality that can be used for prediction.

The usual process here is in several steps. First, the observation of relevant data from which some hypothesis of relationship is derived. Next, predictions are made from the hypothesis and additional data are gathered to see if the predictions are verified. If so, a working tool for making further predictions of the same kind is produced, and the relationship becomes a specific *law* that can be used for this purpose.

If a relationship of this kind is reliable for prediction, but there is no rational explanation of why it works, then it is called an *empirical* law. Because it does seem to work, it can be of use; but a search for a rational explanation is a normal step in the scientific method approach.

All this represents a very powerful process with great practical value, and all executives should be familiar with it and alert to its full potential.

To illustrate this I will go back to early in my career with Media General when I served for a few years as director of research for the company that published the morning, evening, and Sunday newspapers in Richmond. These newspapers had significant coverage in about forty counties of Virginia and we were quite interested, along with our advertisers, in the details of this coverage.

We knew for each county the total circulation of our morning daily and our evening daily, and the household total, but nothing beyond that. Here, for example, might be the information available to us on a given county:

Item	Total	Percent
Households	20,000	100
Morning circulation	10,000	50
Evening circulation	8,000	40

While these data are useful in showing the household coverage of each daily, they leave several questions unanswered for the potential advertiser in both dailies. They reveal nothing about how many households subscribe to both dailies, to either alone, or to neither: all of which is quite relevant to an advertiser. How could they obtain this information?

It was easy enough to pose this question, but not so easy to get the answer for each of the 40 counties involved. A sampling survey would be required for each county and this, with perhaps 300 respondents per county, would involve a total of 300 x 40, or 12,000, respondents altogether: a mammoth and expensive survey undertaking. Further, because of annual change in the basic data, this effort would have to be repeated every year for the finding to remain current. All of this is quite impractical.

Thus we gave up all hope of solving the problem until one day there was a moment of inspiration. The key was always the number of households reading both dailies. If that were known, all the other elements in the puzzle could be derived by simple arithmetic. And the thought occurred that this joint readership might be estimated from the laws of probability.

More specifically, in each case the probability of a household subscribing to each paper is known, and if these events are independent, the law of probability says to multiply these two ratios to find the ratio of both events occurring together. Thus, in the example above, multiply the 50 percent morning coverage by the 40 percent evening coverage to find 20 percent as the expected ratio of households subscribing to both dailies. From this, it is simple to deduce all the elements in the circulation picture as follows:

Item	Total	Percent
Households..........	20,000	100
Subscribers to:..		
Morning only.......	6,000	30
Morning and Evening	4,000	20
Evening only........	4,000	20
Either daily or both	14,000	70
Neither................	6,000	30

The computation process here is quite straightforward. Once you know the joint readership of 20 percent, equal to 4,000 households, simply subtract 4,000 from the morning circulation of 10,000 to find that 6,000 households read the morning paper only. And similarly, when you deduct 4,000 from the evening circulation of 8,000, you find that 4,000 households read the evening paper only. These figures add to 14,000 households reading either daily or both, and when this is subtracted from total households, you find the 6,000 households subscribing to neither daily.

All this seemed much too good to be true. We had great faith in the laws of probability, but for them to work in this case it was necessary that the two events be independent, or very nearly so, and it had simply never occurred to us (or perhaps anyone else involved in newspaper or advertising research) that this might be the case. Indeed, there was good reason to believe the exact contrary to this hypothesis of independence, because the correlation of joint readership and family income was well known. The percentage of households reading two dailies rose steadily with increases in level of income.

However, after this hypothesis was tested with all of the other survey data at hand, it appeared to be valid in each such case. And then it was tested even further against additional surveys and against survey data on newspapers in other areas. Through all of this, the hypothesis emerged unscathed. The multiplicative law of probability for joint events was apparently an excellent predictor of the joint readership of a morning and evening newspaper in a relatively small geographical unit such as an individual county.

As a result, this device was used to estimate the circulation structure in each county and city in the total coverage area, and these estimates were available to advertisers and the circulation department. What would have been an enormous task otherwise came down to nothing more than an hour's work on a calculating machine: a very clear illustration of the power of scientific method in solving problems in the real world.

All of this led to other studies of the coverage of metropolitan dailies throughout the United States, which were widely used in the advertising world.

After all of this became fully apparent, there was a good deal of chagrin at the failure to grasp and test this concept at the very beginning. It seemed quite incredible that we had failed to see, almost at a glance, what now was so simple. However, because it went against all accepted knowledge at the time, it was not so simple or obvious to start with. And this, incidentally, seems often to happen with other projects of this kind. They always seem childishly simple, once the answer is known, but rarely so in the beginning before inspiration strikes.

READERSHIP NORMS

It may be instructive to use another example from this experience, having to do with readership of national advertising in newspapers: a subject of considerable interest to national advertisers and agencies.

Once again, to gather data of this kind was a difficult and expensive process. Interviewers would take a specific edition of the newspaper and go through it page by page with respondents, asking whether they read each item. A great deal of time was involved in this, both in the interviewing process and in the subsequent tabulation of the survey data.

Beyond this, the data from any one readership study had very limited application. The findings were restricted to one edition of a given newspaper and the specific advertisements therein. All this was ill suited to the needs of advertisers and agencies, who were looking for much broader findings of greater generality that could be used in planning future ad campaigns.

At that time a total of nearly 200 studies of this type had been made over the years throughout the country, and readership averages for this entire group were available by type of product, classified separately for men and women. And again, while these data were very interesting, they still lacked the generality needed for predicting the readership to be expected in any given ad campaign.

In looking at the data, it was obvious that readership did not tend to increase in proportion to the size of the ad. If ad size were doubled, for example, you could not expect the percent of readership to double as well. On the contrary, the data suggested that the percent of readership tended to vary with the square root of the advertising space.

Back in those days the typical newspaper page was eight columns wide, and advertising space was measured at 300 lines per column. Thus, a full-page ad was 2400 lines, a half-page was 1200 lines, a quarter-page was 600 lines, and so on. If you take the square root of these numbers you get 49, 35, and 24 respectively, and the readership data seemed to follow percentages similar to this.

The test for this hypothesis consisted of fitting separate straight lines, by least squares, for men and women to the readership data for each of 20 product types, with the independent variable being the square root of the ad size and the dependent variable being the percent of readership. There were 40 such equations, and in each case the equation turned out to be a reasonably close fit.

With the hypothesis thus verified, the square root relationship and the equations derived on that basis could be used with some confidence to predict average readership of any ad of given size for any of the product groups. The equations are shown in Table 24.1, and the expected readership at various size levels is shown in Table 24.2.

All of this was summarized in a brochure that was sent to a mailing list of key advertisers and agencies throughout the nation and for many years was widely used as a planning tool to predict advertising exposure in various campaigns.

Here again, once completed, all this seemed exceedingly simple, but it was not all that obvious at the time. Indeed, this is the only instance I can remember where one variable is related to the square root of another; so it is not too surprising that it took some time to accept and apply this concept.

One thing is unanswered in this investigation. Ad readership clearly seemed to vary with the square root of space, but why should that be so? What rational explanation, if any, could account for this?

178 BASIC BUSINESS STATISTICS FOR MANAGERS

TABLE 24.1
Constants in regression equation
$R = a + b(L)^{1/2}$

Product	N	Men a	Men b	Women a	Women b
Beer and ale	244	−3.2	1.02	−17.2	1.14
Liquor and wines	1574	5.3	0.60	5.1	0.10
Gas and oil	154	6.8	0.63	−5.3	0.54
Passenger cars	717	4.6	0.77	1.3	0.52
Food − total	3395	−0.5	0.22	2.7	0.92
Coffee and tea	399	1.0	0.17	2.8	0.92
Soft drinks	239	−0.4	0.41	3.2	0.76
Cereals	77	−0.4	0.20	0.7	1.04
Condiments	395	0.4	0.14	3.4	0.75
Dairy products	403	−0.8	0.22	3.0	0.92
Meats, fish and poultry	423	−0.7	0.22	3.5	0.95
Soaps and h'sld supplies	729	−0.5	0.23	2.5	0.81
Household equipment	85	−4.6	0.55	3.5	0.71
Financial and insurance	484	−0.5	0.77	−5.5	0.78
Industrial	94	−3.9	1.06	−16.7	1.30
Public utilities	363	7.3	0.34	10.9	0.41
Publications	297	2.7	0.96	16.2	0.64
Tobacco	212	3.5	0.78	−8.5	1.05
Toilet goods	207	4.0	0.23	−3.3	1.00
Transportation	409	3.2	0.64	1.5	.76

There was no definitive answer to this question, but one possible explanation did occur:

The prospect pool for any product is limited. As ad size increases and readers from that pool are attracted accordingly, the remaining pool is diminished; with any further increase in ad size, a constant percentage of that dwindling group will produce a lesser increment of additional readers.

Further, it may be that the visual impact of a given ad is a function of its column width and depth, each of which in a square space is the

TABLE 24.2
PERCENTAGE READERSHIP AT VARIOUS AD SIZE LEVELS

MEN

Product	100	200	400	600	800	1000	1200	1600	2000	2400
Beer and ale	7.0	11.2	17.3	21.9	25.8	29.1	32.2	37.7	45.7	50.1
Liquor and wines	11.3	13.7	17.3	20.0	22.3	24.2	26.0	29.3	32.1	34.7
Gas and oil	13.1	15.7	19.4	22.2	24.6	26.7	28.6	32.0	35.0	37.7
Passenger cars	12.3	15.5	20.1	23.5	26.5	29.0	31.3	35.5	39.2	42.5
Food – total	1.7	2.6	3.9	4.9	5.7	6.5	7.1	8.3	9.3	10.3
Coffee and tea	2.7	3.4	4.4	5.2	5.9	6.4	7.0	7.9	8.7	9.4
Soft drinks	3.7	5.3	7.7	9.6	11.1	12.5	13.7	15.9	17.8	19.5
Cereals	1.6	2.4	3.5	4.4	5.1	5.8	6.4	7.4	8.4	9.2
Condiments	1.8	2.4	3.2	3.8	4.4	4.8	5.2	6.0	6.7	7.3
Dairy products	1.4	2.3	3.6	4.6	5.4	6.2	6.8	8.0	9.0	10.0
Meats, fish and poultry	1.4	2.3	3.6	4.6	5.4	6.1	6.7	7.9	8.9	9.8
Soaps and household supplies	1.8	2.8	4.2	5.2	6.1	6.9	7.6	8.8	9.9	10.9
Household equipment	0.9	3.2	6.4	8.9	11.0	12.8	14.5	17.4	20.0	22.4
Financial and insurance	7.2	10.4	15.0	18.5	21.4	24.0	26.3	30.5	34.1	37.4
Industrial	6.7	11.1	17.3	22.0	26.0	29.5	32.7	38.4	43.4	47.9
Public utilities	10.7	12.1	14.0	15.6	16.8	17.9	19.0	20.8	22.4	23.8
Publications	12.3	16.3	21.9	26.2	29.9	33.1	36.0	41.1	45.7	49.0
Tobacco	11.3	14.5	19.1	22.7	25.6	28.2	30.6	34.8	38.5	41.8
Toilet goods	6.3	7.2	8.6	9.6	10.5	11.2	11.9	13.1	14.2	15.2
Transportation	9.6	12.2	15.9	18.8	21.2	23.3	25.2	28.6	31.6	34.4

Size of Ad in Lines

WOMEN										
Beer and ale	–	–	5.7	10.8	15.1	18.9	22.3	28.5	33.9	38.8
Liquor and wines	6.1	6.6	7.2	7.6	8.0	8.4	8.7	9.3	9.7	10.2
Gas and oil	–	2.3	5.5	8.0	10.0	11.8	13.4	16.3	18.9	21.2
Passenger cars	6.5	8.6	11.6	13.9	15.9	17.6	19.2	21.9	24.4	26.6
Food – total	11.9	15.6	21.1	25.2	28.7	31.7	34.5	39.4	43.7	47.7
Coffee and tea	12.0	15.8	21.2	25.4	28.9	31.9	34.7	39.6	44.0	47.9
Soft drinks	10.8	14.0	18.5	21.9	24.8	27.4	29.7	33.8	37.4	40.7
Cereals	11.1	15.3	21.4	26.1	30.0	33.5	36.6	42.2	47.1	51.5
Condiments	10.9	14.0	18.5	21.9	24.7	27.2	29.5	33.6	37.1	40.3
Dairy products	12.2	16.0	21.5	25.7	29.2	32.2	35.0	40.0	44.3	48.3
Meats, fish and poultry	13.0	16.9	22.5	26.8	30.4	33.5	36.3	41.5	45.9	50.0
Household supplies	10.6	13.9	18.7	22.3	25.4	28.1	30.5	34.9	38.7	42.2
Household equipment	10.6	13.6	17.8	21.0	23.7	26.0	28.2	32.0	35.4	38.4
Financial and insurance	2.3	5.5	10.2	13.7	16.7	19.2	21.6	25.8	29.5	32.9
Industrial	–	1.6	9.2	15.1	20.0	24.3	28.2	35.2	41.3	46.9
Public utilities	15.0	16.7	19.1	21.0	22.6	23.9	25.2	27.4	29.3	31.1
Publications	22.6	25.2	29.0	31.9	34.3	36.4	38.3	41.8	44.8	47.6
Tobacco	2.0	6.3	12.5	17.2	21.2	24.6	27.8	33.5	38.4	42.9
Toilet goods	6.7	10.8	16.7	21.2	25.0	28.3	31.3	36.7	41.4	45.7
Transportation	9.1	12.2	16.7	20.1	23.0	25.5	27.8	31.9	35.5	38.7

square root of the area, or size, of the ad. In this case, readership might well vary with the square root of ad size.

REMAINING PARADOX

One paradox remained in all of this. Readership norms established the square root relationship between ad size and percent of readership. But this raised an interesting question. What this relationship said, in effect, is that a smaller ad is always more efficient than a larger ad in terms of readers per unit of space. And because advertising cost is in direct proportion to space, this meant that the smaller ad is always more efficient in terms of dollar cost per reader.

This being true, the advertiser might well wonder why large space should ever be used: instead of running a full page, why not two half-pages; or instead of a half-page, why not two quarter-pages; and so on. Yet many very astute advertisers, including local department stores and others who keep a close check on sales generated by advertising, were heavy users of full-page and other large-space ads. What was the answer to all this?

This time, a major food chain helped in the search for an answer. Over a period of many months, we ran a series of test ads for this chain, each with coupons from which direct response could be ascertained with great accuracy.

We used the split-run technique in our newspapers in which alternate copies of the same edition go to the mailing room and are delivered to subscribers accordingly. In other words, one household will receive the A version of the newspaper and the household next door the B version, and so on.

Normally these two alternative editions are the same, but they differed during the test. Edition A, for example, might contain one full-page ad and Edition B two half-page ads. The products advertised and the coupons were the same in each case except for size.

There were a number of such tests using various ads that were identical except for size: full-page versus two half-pages, half-page versus two quarter-pages, and so on through a variety of sizes. Then the coupon return was counted in each case. Uniformly, the response from the larger ad was always significantly greater than the combined response from the two half-size ads.

Now, what is the explanation for all this? The conclusion went back to the prospect exposure concept. The most active prospects will search out an ad on a product in which they have a definite buying interest, but the prospect pool is diminished accordingly in terms of response to a second duplicate ad in the same newspaper. The larger ad, by contrast, will attract other prospects with a lesser buying interest, not reached by the smaller ads.

Whether this is the proper explanation or not, there is no question as to the basic findings, which were statistically significant at a very high level. And it appears that the large advertisers were quite correct in their preference for large-space advertising.

In other words, small-space advertising can be very efficient for some advertisers, because it will generate a high rate of prospect readership per dollar of space cost. The prime example is mail order advertising, where relative success can always be measured directly. And, of course, for the same basic reason, there are many others for whom small-space advertising can be highly productive.

But for large advertisers interested in generating a maximum of store traffic or in reaching the maximum audience, full-page ads or the equivalent are clearly the way to get the job done.

SUMMARY

Executives should understand the application of statistics in accordance with the principles of scientific method. This can open the doors to new knowledge and create powerful tools of analysis.

Examples given in this chapter illustrate how the process works and how the testing of one hypothesis can lead to another, with the end result being a far greater understanding of the subject matter as a whole.

CHAPTER 25
Personal Computers

The computer is the marvel of the modern age. Now reaching into almost every facet of our lives, computer technology will undoubtedly continue to develop into even more complex applications well beyond the far reaches of our present imagination.

Initially there was some resistance on the part of executives to personal use of the computer, but this seems to be fading pretty fast as more computers are available to them at home or the office or both.

The simple fact is that a computer is a great time-saving device for executives, up to and including the chief executive officer. It can solve problems in a few seconds that would take far more time even to explain to an assistant, not to mention the wait for the solution to be worked out elsewhere. Thus it is really foolish for any executive, quite irrespective of rank, to pass up a capability of this kind.

PROGRAMMING THE COMPUTER

You don't need to know anything about programming to run a computer any more than you need to be a mechanic to drive a car. Thousands of programs have already been written for you, and with many of these you need do little more than touch a few keys to make them work.

Even so, any reader of this book with access to a personal computer may find it most interesting and instructive to learn a little something about programming. It is very easy to do and much more fun than working crossword puzzles to pass the time on airplane trips.

There are two principal benefits to be derived from this. First, it will give you a much better understanding of what data processing people must do to get the big computers to do their job. And second, you will derive a great deal of satisfaction in being able to direct your own little robot in performing specific tasks.

This can also lead to time saving and greater personal efficiency. Simple programs can be written, for example, to perform all the calculations in the techniques described in this book, requiring nothing more than keying in the original data in each case.

The easiest computer language to learn is BASIC. Although simple to learn and understand, it is nevertheless a very powerful language that can tell the computer to do far more things than you are ever likely to need. There are several versions of BASIC in general use, but all are quite similar in structure and key commands.

Anything that varies is known to the computer as a variable, and in BASIC you can give this any name that does not exceed eight characters, such as A or X or SALES or PROFIT. If it is a *string* expression involving letters or words, then a $ sign is added to the variable name, such as A$, X$, and so on.

Each separate instruction to the computer is in a line called a *statement*. These are numbered, and the computer will always execute them in their numerical order. It is customary to number the statements in intervals of 10, such as 10, 20, 30, and so on. This permits the later interspersing, if required, of other statements with numbers like 15 or 22. But the computer cares nothing about this and will execute the statements in strict numerical sequence no matter what the intervals between them may be.

Symbols used to tell the computer which arithmetical procedure to follow are these:

Symbol	Example	Means
+	A + B	A plus B
−	A − B	A minus B
*	A * B	A times B
^	A ^ B	A to the B power
=	A = B	A equal B
>	A > B	A greater than B
<	A < B	A less than B

All of this is very straightforward and easy to remember.

There are three principal ways for the computer to access data to work with. One is to read data from a separate file that is stored on a disc. A second is to ask the user to supply the data when the program runs. And a third way is through a DATA statement in the program itself.

The computer asks the user to supply data through an INPUT statement, as follows:

10 INPUT X

or

10 INPUT "X...", X

When the computer comes to the first statement, it will display a question mark (?) and then wait for entry of the number to be keyed in that is to be assigned to the variable X. With the second statement, it will display "X..." and then wait for the number to be entered. This is often helpful in prompting the user as to the proper number to enter.

A DATA statement takes the form:

10 DATA 10,22,47,58

When directed to do so, the computer will read these numbers in sequence and work with them accordingly. It will read all DATA statements in this fashion, in numerical order, no matter where they appear in the program.

Thus there are some very simple and easy ways to feed data into a program. This can be illustrated with an example used elsewhere in this book, where a bank has a monthly service charge for certain checking accounts equal to $5 plus 10 cents per check:

```
10 INPUT "NUMBER OF CHECKS...", NUMBER
20 CHARGE = 5 + .10*NUMBER
30 PRINT "SERVICE CHARGE..."; CHARGE
   RUN
NUMBER OF CHECKS...15
SERVICE CHARGE...     6.50
```

This little program will print out the service charge for any number of checks entered when the computer pauses after statement 10.

Now, suppose the bank decides to permit five free checks a month. How can the program be modified to handle this? Just add this further statement:

25 IF NUMBER < 6 THEN CHARGE = 5

This has the effect of overriding line 20 and thereby eliminating any surcharge for the first five checks. This is known as a *contingent* statement, and the use of such statements makes it possible to create many kinds of program logic to control decisions made by the computer.

The statement above, incidentally, can be added at the end of the original program and the computer will automatically put it in proper sequence. You can see this at any time by entering the LIST command, whereupon the computer will list the entire program with all statements in proper numerical sequence.

Now, see how a program can feed data into the computer through a DATA statement. Here is a simple program to sum three numbers:

```
10 DATA 10,20,30,0
20 READ X
30 IF X = 0 THEN 100
40 SUM = SUM + X
50 GOTO 20
100 PRINT "SUM + ";SUM
    RUN
SUM = 60
```

The DATA statement is in line 10. Add a zero at the end of the three numbers to be used as a trigger to tell the computer (in line 30) that there are no more numbers to be read.

In line 20, the computer is told to read the numbers in the DATA statement in serial order, one by one, and to assign each such number to a variable named X.

In line 30, the computer is told to skip on to line 100 if the number read is zero and to complete the program. To trigger this action any other number could have been chosen, with an appropriate change, of course, in line 30.

Line 40 does the work of adding these various numbers. It may look like algebraic nonsense, but the = sign in BASIC has a different meaning from that in algebra. In BASIC, the = sign means that the value on the right is assigned to the variable on the left, which is a very versatile and useful strategem.

The statement SUM = SUM + X means in this case that every time the computer reads a number from the DATA statement, it is added and accumulated in the variable designated as SUM. And this value is finally printed out by the PRINT statement in line 100.

Please also note line 50. After adding each number read to SUM, the computer with the GOTO 20 statement is told to go back to line 20 and repeat the process.

In effect, a looping procedure has been set up that repeats itself over and over until terminated by the zero in the DATA statement. This looping procedure is used in many ways in computer programming. It is a very powerful device in solving a great variety of problems.

Finally, note the apostrophe before the variable name in the PRINT statement. As you might expect, the computer is quite inflexible in matters of punctuation, and nothing but an apostrophe will be accepted in a PRINT statement of this kind. Try a comma instead and you will be admonished accordingly by a stern ERROR message!

Now, to illustrate another looping procedure, find the present value of a stream of payments for the next five years where the assumed interest rate is 10 percent. This little program will do the job:

```
10 DATA 10000,12000,15000,19000,24000
20 FOR K = 1 TO 5
30     READ X
40     PVAL = X/1.1 ^ K
50     SUM = SUM + X
60     SUM PV = SUM PV + PVAL
70 NEXT K
80 PRINT "DOLLAR PAYMENTS..."; SUM
90 PRINT "PRESENT VALUE..."; SUMPV
   RUN
DOLLAR PAYMENTS...80000
PRESENT VALUE...   58157.35
```

Annual payments are listed from Year one to Year five in the DATA statement in line 10, and the computer reads these numbers, one at a time, and assigns them to the variable X.

Line 20 begins the looping procedure in this program. The computer will start with the first value of K, or one in this case, and will then process all succeeding statements through line 60. Line 70 will than direct it to return to line 20 and repeat the process through K=5. After this, the computer will move on to line 80 and complete the program.

Here is another example. Suppose you want to know the value of one dollar, compounded at the rate of 10 percent, at the end of each year for ten years. The small program in Table 25.1 does the job, but two things should be noted about the printed output:
1. The numbers are printed in two columns that are 15 spaces wide. This will always occur when the variables in the PRINT statement are separated by a comma. If you had used a semicolon instead, the two columns would have been printed closer together, only one space apart.
2. In the second column, the computer prints all the decimal numbers it can in each case, and it will always do this unless instructed otherwise. The result, as in this instance, can be a very ragged column of figures. You can change all this by adding a new statement (line 5) to the program and by slightly amending line 30, as is shown in Table 25.2.

Here the format is set in line 5 with the string expression assigned to the variable A$. Each "#" symbol assigns one space per digit, and a decimal point can be placed wherever you wish.

In this example, there are six spaces for a whole number in the first column and space for eight digits before and two digits after a decimal point in the second column. The PRINT statement in line 30 carries out these instructions accordingly, with a much tidier result.

TABLE 25.1

10 FOR YEAR = 1 to 10
20   VALUE = 1.1^YEAR
30   PRINT YEAR, VALUE
40 NEXT YEAR

1	1.1
2	1.21
3	1.331
4	1.4641
5	1.61051
6	1.771561
7	1.948717
8	2.143589
9	2.357948
10	2.593743

TABLE 25.2

```
5  A$ = "###### #######.##"
10 FOR YEAR = 1 to 10
20     VALUE = 1.1^YEAR
30     PRINT USING A$; YEAR, VALUE
40 NEXT YEAR
```

1	1.10
2	1.21
3	1.33
4	1.46
5	1.61
6	1.77
7	1.95
8	2.14
9	2.36
10	2.59

Where this is done, incidentally, the computer always rounds up the last digit shown if the next number is five or more.

So, it is very easy to format the output of any computer program in terms of column spacing and decimal numbers to display the results exactly as you want.

Now, here is a final example to illustrate all this. If you travel overseas, you will hear temperatures given in centigrade. This is a little difficult to translate into the kind of degrees you are familiar with. The precise relationship is:

$$\text{Fahrenheit} = 1.8(\text{Celsius}) + 32$$

This is not very easy to work out in your head. So perhaps there is another easier formula that will give a fair approximation to the real answer. Such a one might simply be to double the Celsius figure and add 28. The question is how accurate this is, and the computer works this out in Table 25.3. Clearly, since the small estimating formula comes fairly close to the real answer over the specified range of temperatures, it is a handy thing to remember on your next trip abroad.

TABLE 25.3

```
 5 PRINT:PRINT:PRINT
10 A$="########     ########.#     ########.#     ########.#"
20 PRINT "    CENT         FAHR            EST            ERROR"
30 PRINT
40 FOR CENT = 10 to 30
50   FAHR = 1.8*CENT + 32
60   EST  = 2*CENT + 28
70   DEV  = EST - FAHR
80   PRINT USING A$; CENT, FAHR, EST, DEV
90   IF CENT = 15 OR CENT = 20 OR CENT = 25 THEN PRINT
100 NEXT CENT
```

Cent.	Fahr.	Est.	Error
10	50.0	48.0	−2.0
11	51.8	50.0	−1.8
12	53.6	52.0	−1.6
13	55.4	54.0	−1.4
14	57.2	56.0	−1.2
15	59.0	58.0	−1.0
16	60.8	60.0	−0.8
17	62.6	62.0	−0.6
18	64.4	64.0	−0.4
19	66.2	66.0	−0.2
20	68.0	68.0	0.0
21	69.8	70.0	0.2
22	71.6	72.0	0.4
23	73.4	74.0	0.6
24	75.2	76.0	0.8
25	77.0	78.0	1.0
26	78.8	80.0	1.2
27	80.6	82.0	1.4
28	82.4	84.0	1.6
29	84.2	86.0	1.8
30	86.0	88.0	2.0

All the statements in the program should be quite familiar by now except line 5, which causes the computer to space three lines before printing the results. Similarly, the PRINT statement in line 30 creates a skipped line after the column heads are printed, and line 90 creates a blank line every fifth line to make the table easier to read.

This is a trivial example, but it does illustrate how easily the computer can deal with problems of this kind.

SUMMARY

The computer is a tremendous help and a great problem solver in the modern world, and its importance is certain to grow in the years ahead. Executives should learn how to take full advantage of this remarkable machine and, although it is not essential, will find it very useful to learn at least the fundamentals of the BASIC language so that they can communicate directly with the computer.

This is a very easy thing to do. Indeed, if you have learned the simple instructions in this chapter, you already know how to program the computer to solve a wide variety of problems. And there are many books available that are easy to understand and can add to your knowledge of BASIC. None of this is a chore. It really is great fun and should be enjoyed as such by any executive, irrespective of rank and position!

APPENDIX

Exercises

FREQUENCY DISTRIBUTIONS

Exercise 1:
The following figures represent the percent of quota various branches of a regional bank achieved in opening new business accounts. How would you summarize these data for presentation to top officers of the bank?

104	139	116	127	78
72	112	85	51	149
85	90	107	112	118
106	72	53	117	86
68	122	82	90	118
105	74	77	58	91
105	60	142	127	51
66	67	117	122	109
114	149	82	75	77
97	52	91	129	114

Answer:
The data can be most conveniently summarized in a frequency distribution, as follows:

From	To	Frequency
50	60	5
60	70	4
70	80	7
80	90	5
90	100	5
100	110	6
110	120	9
120	130	5
130	140	1
140	150	3
Total		50

Exercise 2:
Compute the arithmetic average of the data summarized in the following frequency distribution:

Class	Frequency
10 to 20	5
20 to 30	20
30 to 40	12
40 to 50	8
50 to 60	5
Total	50

Answer:
The arithmetic mean is 32.6. This is computed using the midpoint of each class, 15, 25, and so on, which is multiplied by the frequency involved in each case. The sum of these products is then divided by the number of items (50) to get the arithmetic mean.

AVERAGES

Exercise 1:
Weekly salaries of a group of employees are listed below. How does the arithmetic mean of the series compare with the median, and which is the more representative of the entire group?

425	915	615
750	525	1,725
600	1,575	805

Answer:
The arithmetic mean is 881.7 and the median is 750. The latter is more representative of the group, as it usually tends to be with data of this type.

Exercise 2:
If an accounting team completes one-half of the accounts being verified at the rate of 8 per labor-hour and the second half at 12 per labor-hour, what is the average for the entire job, and which mean is used to determine this answer?

Answer:
The answer is 9.6 accounts per labor-hour and is determined by the harmonic mean. The arithmetic mean of 10 is incorrect in this context.

Exercise 3:
If the sales of your division, expressed as a ratio of the preceding year in each case, have been 80, 110, and 140 percent respectively in the last three years, what has been the average rate of change, and which mean gives you the correct answer?

Answer:
The correct average is 107.2, indicating an average increase of 7.2 percent per year. It is determined by the geometric mean. The arithmetic mean of 110, indicating a 10 percent average increase, is incorrect.

You can check this out by starting with any base number, such as 1,000, and reproducing the sales figures accordingly (1,000, 800, 880, and 1,232). Now, if you apply the 110 ratio each year, the final total is an incorrect 1,331; whereas applying 107.2 percent each year produces the proper ending total of 1,232.

WEIGHTED AVERAGES

Exercise 1:
Here are the percentages of overdue accounts in three categories of accounts receivable. What is the average percent for the entire group?

Category	Accounts	Pct. overdue
A	840	6.1
B	1,770	8.2
C	4,130	12.7
Total	6,740	?

Answer:
The correct figure is the weighted average of 10.7 percent, where each individual percent is weighted by the number of accounts involved. This contrasts with the incorrect figure of 9.0 percent based on the unweighted average of the three percentages.

Exercise 2:

Here are the hours of operation and the output per hour of two machines for two consecutive weeks. What is the average output per hour in Week one versus Week two for both machines combined?

Machine	Hours		Output per hour	
	Wk. 1	Wk. 2	Wk. 1	Wk. 2
A	80	40	100	105
B	40	80	60	65
Total	120	120	?	?

Answer:

Weighted by the number of hours in each case, the combined average output was 86.7 per hour the first week versus 78.3 the second week.

Exercise 3:

How much did the change in mix contribute to the decline in output per hour for the two machines combined?

Answer:

If the hours of operation of each machine had been the same in Week two as in Week one, average output per hour would have increased from 86.7 to 91.7. The difference between the latter figure and the 78.3 weighted average is attributable to the change in mix.

VARIABILITY

Exercise 1:

Why is the range in a series of data, the difference between the lowest and the highest value, a relatively poor measure of variability?

Answer:

It is based on only two figures and reveals nothing about the remainder of the series.

Exercise 2:

What does it mean if a number exceeds the third decile in its series?

Answer:

The number is in the top 30 percent of the series, exceeding at least 70 percent of all the others.

Appendix **197**

Exercise 3:
What does it mean if a number is in the top quartile of its series?

Answer:
It is in the top one-fourth of the series, exceeding at least 75 percent of all the others.

Exercise 4:
What does the 85th percentile of a series mean?

Answer:
It means that 85 percent of the numbers are lower and 15 percent are higher than the specified value.

Exercise 5:
Compute the average deviation and the standard deviation in the following series:

$$2$$
$$4$$
$$6$$
$$8$$
$$10$$

Answer:
The average deviation is 2.4, and the standard deviation is 2.828.

Exercise 6:
In the numbers just given, what is the short-cut method of calculating the sum of the squared deviations?

Answer:
Add up and square the original numbers, and derive the sum of squared deviations from Sum $(X)^2$ − (Sum $X)^2/N$ = $220 - 30^2/5$ = $220 - 180 = 40$.

Exercise 7:
How do you compare the relative variation in these two series of numbers?

	A	B
Mean..................	20	40
Standard deviation	3	4

198 BASIC BUSINESS STATISTICS FOR MANAGERS

Answer:
The coefficient of variation is 15 percent in the A series versus 10 percent in the B series.

RATIOS AND PERCENTAGES

Exercise 1:
What is the difference between 12.7 percent and 127 per 1000?

Answer:
No difference. The value of a ratio is not changed by shifting the base, and in this case the base is merely shifted from 100 to 1000.

Exercise 2:
Here your company's revenue is shown along with the gross national product. What ratio would you use to show the relationship between the two?

Year	GNP ($Billions)	Revenue ($Millions)
0	3,765	123
1	3,998	152
2	4,256	197

Answer:
A convenient base here is $1 million of GNP, which yields the following comparison:

Year	GNP ($Billions)	Revenue ($Millions)	Per $Million GNP
0	3,765	123	32.67
1	3,998	152	38.02
2	4,256	197	46.29

Exercise 3:
With data classified by rows and columns, how many ways can you compute percentages?

Answer:
Three ways, and in each case row totals, column totals, and the grand total equal 100 percent. Which to choose, or whether to choose

them all, depends on the nature of the data and use for which it is designed.

COMPOUNDING

Exercise 1:
If you invest $1,000 in a savings account to earn interest at the rate of 10 percent per annum, what value will be accumulated in one year if interest is compounded: (A) annually, (B) daily?

Answer:
With annual compounding, the savings account will grow to $1,100; with daily compounding, $1,105.16.

Exercise 2:
If $10 had been invested 200 years ago at 10 percent per annum, what would be the current value of the investment?

Answer:
Current value would be $1.9 billion.

Exercise 3:
If in the last five years your division revenue has grown from $276,000 to $834,000, what has been the annual compound rate of increase?

Answer:
Compound annual increase of 25 percent.

PRESENT VALUE ANALYSIS

Exercise 1:
What is a dollar payable in ten years worth today if the time value of money is 10 percent per annum?

Answer:
Present value is 38.55 cents.

Exercise 2:
What is this same dollar worth if the time value of money is 6 percent per annum?

Answer:
Present value is 55.84 cents.

Exercise 3:

An investment opportunity is made available whereby you invest $15,000 and then get the return indicated below at the end of each specified year in the future. If the time value of money is 10 percent per annum, is this a good investment?

Year	Return
6	$5,000
7	5,000
8	5,000
9	5,000
10	5,000
11	5,000
Total	30,000

Answer:

Not a good investment. This future stream of payments has a present value of $13,521, which is less than the amount being invested.

Exercise 4:

What fact is often omitted when executives are given the result of a present value analysis?

Answer:

Quite often, those who make the analysis neglect to mention the interest rate that was used in their calculation. This, of course, is a critical element and should always be specified accordingly.

INTERNAL RATE OF RETURN

Exercise 1:

What is the special value of an internal rate of return analysis?

Answer:

The internal rate of return has the great virtue of relating the return on a given investment, in the form of a stream of future payments or benefits, to the cost of the investment. Furthermore, it accomplishes this complex analysis through a single abstract ratio that is fully comparable from one investment analysis to another. It thus fills a vital role in the decision-making process in the business world.

Exercise 2:

When it is reported that a given investment has a rate of return of some given figure, say 15 percent, what precisely does this mean?

Answer:

It means that if all future payments or benefits are discounted by this rate of return, the resulting present value is exactly equal to the initial cost of the investment. In other words, the investment is returned even if all future dollars are discounted at the specified return rate.

Exercise 3:

Suppose the chief executive officer of your company has established a rule that no investment will be approved unless the expected return is at least 15 percent.

The computer division has requested approval for a $100,000 expenditure, for which dollar savings are projected below. Does this project meet the minimum return criterion?

Year	Return ($000)
1	10
2	10
3	10
4	20
5	20
6	20
7	30
8	30
9	30
10	40
Total	220

Answer:

It does not meet the minimum criterion. The rate of return is only 13.4 percent.

Exercise 4:

Your organization has an opportunity to acquire another company for $1 million. The new company is expected to earn $100,000 after taxes the first year after acquisition, and this is projected to increase by $20,000 a year for the next ten years. How would you calculate the rate of return on this investment?

Answer:

You must first fix upon some terminal value of investment, and it is fairly common in this situation to assume that the acquired company will then have a going value equal to ten times its net income in the prior year. On that basis, you would calculate a 20.9 percent rate of return, as follows:

Investment 1,000
Rate of return 20.89%

Year	Value of $1	Annual return	Present value	Cumulative
1	0.8272	100	83	83
2	0.6842	120	82	165
3	0.5660	140	79	244
4	0.4681	160	75	319
5	0.3872	180	70	389
6	0.3203	200	64	453
7	0.2649	220	58	511
8	0.2192	240	53	564
9	0.1813	260	47	611
10	0.1499	280	42	653
11	0.1240	2,800	347	1,000
Total		4,700	1,000	

PROBABILITY

Exercise 1:

In considering a new business venture, you can identify two mutually exclusive contingencies that would cause it to fail. You evaluate each contingency as having a likelihood of 20 percent. What is the probability that the venture will succeed?

Answer:

There is a 40 percent chance of failure and a 60 percent chance of success, based on the addition law of probabilities.

Exercise 2:

Assume in the preceding example that both events must occur for the venture to fail. What then is the probability of success?

Appendix **203**

Answer:
There is a 96 percent probability of success, based on the multiplication law of probabilities.

Exercise 3:
There are ten critical parts in a machine and all must function correctly for an operation to succeed. Each has an expected failure rate of 1 in 100. What is the probability that the operation will succeed?

Answer:
The probability of success is 90.4 percent (.99 to the tenth power) and the probability of failure is 9.6 percent.

Exercise 4:
After ten consecutive flips of a coin with heads the result each time, what is the probability of another head on the next flip?

Answer:
If the coin is well balanced, the probability of another head continues to be 50 percent on the eleventh or any other succeeding flip of the coin.

PERMUTATIONS AND COMBINATIONS

Exercise 1:
What value does 5! represent?

Answer:
This is 5 factorial (5 × 4 × 3 × 2 × 1), which has the value of 120.

Exercise 2:
What is 8!/6!?

Answer:
This is 8 × 7 = 56.

Exercise 3:
How do you express 1.23 billion in scientific notation?

Answer:
As 1.23E9.

Exercise 4:
How many ways can you arrange ten different things taken four at a time?

Answer:
A total of 5,040 ways.

Exercise 5:
How many teams of four each can you select from a pool of ten executives?

Answer:
A total of 210 different teams.

STATISTICAL INFERENCE

Exercise 1:
How is it possible to know how statistics are likely to vary, purely as a matter of random sampling?

Answer:
Many such statistics are normally distributed, or very nearly so, and the details of the normal distribution are based on a mathematical formula and are precisely known.

Exercise 2:
What two parameters determine how all data items in a given series will be distributed in a normal distribution?

Answer:
The arithmetic mean and the standard deviation of the series.

Exercise 3:
What is a normal deviate?

Answer:
It is the deviation of a given item from the arithmetic mean divided by the standard deviation. In other words, it is the deviate expressed in standard deviation units. Mathematical tables on the normal distribution are based on deviates in this form.

Exercise 4:
In a normal distribution, what percentage of the data items will be included in the range of the mean plus and minus the standard deviation?

Answer:
Approximately two-thirds of the data items will be included in this range, and one-third will lie outside.

Exercise 5:
In a normal distribution, how many data items will lie outside the range of the mean plus and minus two times the standard deviation?

Answer:
About 5 percent of the items will lie outside this range.

Exercise 6:
How many data items will lie outside the range of the mean plus and minus 2.8 times the standard deviation?

Answer:
About 1 percent of the items in a normal distribution will lie outside this range.

Exercise 7:
What does it mean to say that a given statistic is significant at the .05 level?

Answer:
That the probability of such a statistic occurring purely as a matter of chance due to random sampling is 5 percent or less.

Exercise 8:
What is meant by a null hypothesis?

Answer:
A null hypothesis specifies a certain mean or ratio or other statistic in the population being studied. A sample is drawn from the population to test the hypothesis. The hypothesis is rejected if the equivalent sample statistic lies outside a preset reliability limit, such as the .05 level of probability, in terms of deviation from the assumed population value.

INTERPRETING SURVEY RESULTS

Exercise 1:
In a properly selected sample of 100 customers, you find that the average number of orders placed in a given period was ten, with a standard deviation of three. How much reliance can you place in this average?

Answer:
The standard error of the mean is .3, so you can infer that the odds are about two in three that the population mean is in the range

of 10 +/− .3 and about nineteen in twenty that it is in the range of 10 +/− .6.

Exercise 2:

In a marketing survey of consumer preference, a properly selected sample of 100 shows a preference ratio of 30 percent. How much confidence can you place in this percentage? What if the sample size had been doubled to 200?

Answer:

With a sample size of 100, the standard error is 4.58 and twice that is 9.16, so you can conclude that the odds are about nineteen in twenty that the population mean is in the range of 30 percent +/− 9.2. With a sample size of 200, the same odds would apply to a range of 30 percent +/− 6.5.

Exercise 3:

In testing employee preference for a given benefit plan, with a sample size of 100 in each case, 40 percent of the respondents in one division favored the plan versus 30 percent in another division. Can this be taken as conclusive proof of a real difference of opinion between the two groups of employees?

Answer:

The standard error of the difference is 6.7, and the sample difference of ten is only 1.49 times this standard error. Normally, any deviation less than twice the standard error would not be considered as statistically significant. If further evidence is needed on this point, a larger sample could be taken.

SAMPLING TECHNIQUES

Exercise 1:

If random sampling is required to draw reliable inferences about the whole population from a sample, what does *random* mean in this context? Does it mean that the sample is picked purely by happenstance?

Answer:

No. Quite the opposite of happenstance. It means that each member of the population being sampled, or a defined segment thereof, has an equal probability of being included in the sample.

Appendix   **207**

Exercise 2:

In an effort to measure public opinion in a given city, you select respondents at random from those at a given street intersection during the normal lunch hour period. Is this a valid sample for your purposes?

Answer:

No. There is no way you can project the findings in such a poll to the population of the city with any known degree of reliability.

Exercise 3:

You want to select a random sample of 100 employees from a company total of 1,000. Can you do this by simply picking 100 names by happenstance from the company roster of employees?

Answer:

This is not a satisfactory way of picking a random sample because there is no assurance that each employee would have an equal chance of being selected.

Exercise 4:

You want to pick a sample of customers to analyze. Is a pure random sample the most efficient way to approach this?

Answer:

It may be more efficient to divide the customers into groups based on size or some other criterion and sample within each group. The sampling rate can vary from one group to another, in accordance with its variability. Within the large customer group, for example, the sampling rate could be 100 percent. Final results, of course, would be weighted accordingly.

Exercise 5:

If a mail survey claims a 60 percent response rate, does that mean its findings are valid?

Answer:

While 60 percent responded, the survey reveals nothing at all about the 40 percent who did not respond.

Exercise 6:

How is it possible to correct for the nonresponse ratio in a sample survey?

Answer:
Those who did not respond can be interviewed again, and the entire group can be treated as a separate category with sampling results weighted and projected accordingly.

Exercise 7:
In terms of reliability, are larger samples needed for larger populations?

Answer:
Unless the sample is a high percentage of the population, which is rarely the case, the reliability of a sample is a function of sample size only, and the size of the population being sampled is essentially irrelevant.

CHI SQUARE

Exercise 1:
A marketing survey sample reported consumers of two competitive products as follows:

Women	Consumers of	
	A	B
Under 35	121	175
35 & over	190	163

How many degrees of freedom are involved in this fourfold table, and why?

Answer:
Only one degree of freedom because, with row and column totals fixed, any change in one cell automatically adjusts the other cells accordingly.

Exercise 2:
What does chi square reveal about the significance of the differences reported in product consumption by women in the two age groups?

Answer:
Chi square is 10.81, which compares with a value of 6.64 at the .01 level. Thus, the deviation in product consumption by age group

would occur far less than 1 percent of the time as a matter of chance and is highly significant from a statistical viewpoint.

Exercise 3:
Participating in a company thrift plan that involves company stock is often viewed as an index of relative morale. Such participation by various divisions of your company is shown below. Can you conclude that the different divisions differ significantly in their rate of participation?

Division	Total	Participants	Percent
A	225	132	59
B	183	81	44
C	261	128	49
D	341	221	65
E	162	98	61
Total	1,172	660	56

Answer:
There are five degrees of freedom here, one for each division, and the value of chi square is 28.07, which is far greater than the .01 value of 15.09. Thus, the differences are highly significant from a statistical viewpoint.

Details of the chi square calculation are as follows:

Data

Category	Group A	Group B	Total
1	132	93	225
2	81	102	183
3	128	133	261
4	221	120	341
5	98	64	162
Total	660	512	1172

Chi square calculation

Group	Theo.	Actual	Diff.	Diff.²	Chi sq.
A	126.7	132	5.3	28.02	0.22
B	98.3	93	−5.3	28.02	0.29
A	103.1	81	−22.1	486.41	4.72
B	79.9	102	22.1	486.41	6.08
A	147.0	128	−19.0	360.22	2.45
B	114.0	133	19.0	360.22	3.16
A	192.0	221	29.0	839.22	4.37
B	149.0	120	−29.0	839.22	5.63
A	91.2	98	6.8	45.85	0.50
B	70.8	64	−6.8	45.85	0.65
Chi square...					28.07

(Note: Computer calculated prior to rounding of theoretical number.)

ANALYSIS OF VARIANCE

Exercise 1:

Three different machines are tested with five operators and give the following results in terms of units produced in a specified period of time:

	Operator			
Machine	A	B	C	Mean
1	212	230	196	212.7
2	196	212	188	198.7
3	195	223	195	204.3
4	192	202	179	191.0
5	212	220	193	208.3
Mean	201.4	217.4	190.2	203.0

Are these mean differences by machine and by operator statistically significant?

Answer:
There is a significant difference in variance by operator and by machine based on an F value that far exceeds the .05 level of probability in each case. The analysis is as follows:

Source of Variation	Degrees of freedom	Sum of squares	Mean square	F
Rows	4	859	214.8	9.42
Columns	2	1869	934.5	40.98
Remainder	8	182	22.8	-
Total	14	2910	207.9	-

REGRESSION AND CORRELATION

Exercise 1:
If a telephone company charges a fee of $10 per month plus 20 cents for each call made, how would you write this in equation form?

Answer:
Monthly charge = 10 + .20 (number of calls).

Exercise 2:
In this equation, normally written as $Y = a + bX$, what is the coefficient a called and what does it mean?

Answer:
It is called the *intercept* because it intercepts the Y scale and represents the Y value accordingly when X is zero. In this example, the monthly charge is $10 even when no calls are made.

Exercise 3:
What is the coefficient b called and what does it mean?

Answer:
This is the coefficient of regression, and it represents the change in Y for every unit change in X. In this example, the monthly charge increases 20 cents for every call made.

Exercise 4:
In plotting Y versus X on a chart, which axis is used for each?

Answer:
Y is the dependent variable and is plotted on the vertical axis. It depends on X, which is the independent variable plotted on the horizontal axis.

Exercise 5:
When you plot values in the Y = a + bX equation, what sort of line do you get and how many values to you need to plot the line?

Answer:
A straight line that can be plotted correctly from any two points on the line.

Exercise 6:
In studying a relationship between two variables, what is normally the first step in the analysis?

Answer:
It is normally desirable to plot the data in a scattergram, where a dot or circle represents the intersecting point of each pair of numbers. Visual inspection is then usually sufficient to decide whether a straight line can properly represent the relationship.

Exercise 7:
When a straight line is fitted to two variables by the least squares method, what is the result?

Answer:
The straight line thus derived best fits the data in the sense that it minimizes the sum of the squared deviations from the line.

Exercise 8:
What is meant by the standard error of estimate, the coefficient of determination, and the coefficient of correlation, as derived in a least squares analysis?

Answer:
The standard error of estimate is a measure of dispersion around the fitted line. Typically, like the standard deviation, it will include in its range about two-thirds of all the deviations from the line.

The coefficient of determination can be interpreted as the ratio of variation in the dependent variable that is associated with or explained by the least squares relationship. The coefficient of correlation is the square root of the coefficient of determination. Both of these are abstract measures and thus are fully comparable with others of the same type.

Exercise 9:

The following data show the percent of overtime in a manufacturing plant for a period of ten consecutive weeks and the percent of units produced that were found to be defective by the quality control section. Do these data indicate a relationship between these two variables, and, if so, how can it be represented?

Week	Percent overtime	Defective rate
1	14.2	3.4
2	7.5	2.3
3	17.8	3.9
4	19.5	4.2
5	1.0	1.0
6	2.1	1.7
7	16.3	3.9
8	20.8	4.8
9	11.6	2.8
10	5.8	2.0

Answer:

Plotting the data makes it clear that a straight line relationship is appropriate. A least squares analysis produces the equation involved, as follows:

$$Y = 1.026 + .169x$$

Here .979 is the coefficient of determination, indicating that about 98 percent of the variation in the defective rate is explained by variation in the overtime rate: a very strong relationship indeed.

The final conclusion here is that the cost involved in the defective product rate should be considered in calculating the total cost of overtime in the plant operation. (See Figure A-1.)

TREND ANALYSIS

Exercise 1:

You are asked to make a three-year forecast of your industry sales, shown below in millions of dollars. What would be your forecast based on a straight line fitted to the data by least squares?

OVERTIME AND PRODUCT DEFECTIVE RATE
Based on Data for 10 Consecutive Weeks

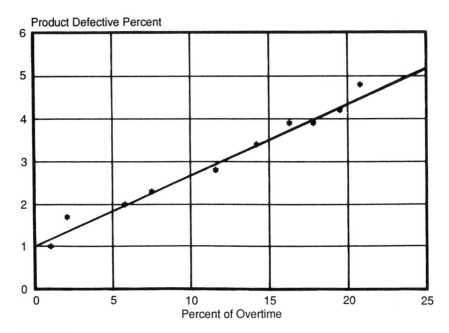

FIGURE A-1

Year	Sales
1	557
2	803
3	956
4	1,121
5	1,483

Answer:

Extrapolated values of the equation are:

Year	Sales
6	1,635
7	1,852
8	2,069

Exercise 2:

How well does the equation fit the data?

Answer:

The coefficient of determination is .973, indicating that more than 97 percent of the variation in sales is explained by the equation; the standard error of estimate is 65.2 million dollars.

This indicates that the relationship represented by the equation is a good fit, but this applies only to the fitted data themselves and not to the extrapolation. Reliability to be placed in the latter is judgmental in nature.

Exercise 3:

Based on the trend analysis, what were the relatively good and bad years for the industry in the past five years?

Answer:

Both Years two and five were 5 percent over trend, and Year one was 1 percent over. Year three was 3 percent under and Year four was 7 percent under.

SEASONAL VARIATION

Exercise 1:

What is the error potential in the usual practice of dealing with seasonal variation by comparing data with the same period in prior years?

Answer:

The natural tendency to interpret such changes wholly in terms of the current year, thus ignoring any abnormal influences in the same period of the prior year.

Exercise 2:

What is the prime advantage of adjusting data for seasonal variation?

Answer:

Such data can be compared with that of any other period and are not restricted to comparison with the same period of prior years. The result is a far clearer picture of underlying trends.

Exercise 3:

What is the basic concept involved in the derivation of seasonal indexes? Use monthly indexes as an example.

216 BASIC BUSINESS STATISTICS FOR MANAGERS

Answer:
The object is to find the typical relationship of each month to the year as a whole, or to the average month of that year.

Exercise 4:
What is the actual process involved in deriving seasonal indexes for a given set of data?

Answer:
Over a period of years in the past, the percent of each month to its centered monthly average for the year is calculated, and the arithmetic mean of these percentages is then computed accordingly. These means are then adjusted up or down by a common factor to insure that they add to 1200 for all twelve months combined.

Exercise 5:
Once the seasonal indexes are obtained, how are the actual data adjusted to eliminate seasonal variation?

Answer:
All that is necessary is to divide the actual data for each month by its seasonal index in ratio form. The result is the adjusted data, fully comparable for any period versus another.

Exercise 6:
Is there any other way to present seasonally adjusted data, and what advantage does this have?

Answer:
The adjusted data can be restated in index form, with the total for any selected year, or average for several years, made equal to 100. All the data are then simply shown as a percentage of this total or average.

The advantage of this is that the index numbers are often easier to read and digest than the adjusted numbers; and because they are abstract in nature, they can easily be compared with other time series.

Exercise 7:
Here are the sales of a given product in the four quarters of the prior year. Seasonal indexes for Quarters one through four are 1.35, 1.26, and .83, and .56 respectively. What is the underlying trend in these data?

Quarter	$000
1	1,640
2	1,760
3	1,247
4	1,016

Answer:
Adjusted for seasonal variation, the data are as follows:

Quarter	$000
1	1,215
2	1,397
3	1,502
4	1,814

The strong upward trend is quite evident.

COMPOUND GROWTH RATES

Exercise 1:
Why is it useful to compute the compound growth rate in a given time series?

Answer:
Because this is an abstract measure and thus comparable to similar rates computed for other series.

Exercise 2:
What is wrong about using the first and last number in a series to compute the compound growth rate?

Answer:
Such a rate is based only on the two years and ignores everything in between. As such, it can be a very poor and misleading average for the series as a whole.

Exercise 3:
Use the proper method to find the compound annual growth rate in the following series.

Year	$000
1	1,127
2	1,346
3	1,595
4	1,810
5	2,021
6	2,276

Answer:
Fitting a straight line to the natural logs of the data produces a compound growth rate of 14.9 percent, with a consistency ratio (coefficient of determination) of 98.9 percent.

The equation is:

$$\text{Log}Y = 7.408 + .1389(x)$$

The antilog of the regression coefficient is 1.149, equal to the 14.9 percent compound rate.

INDEX NUMBERS

Exercise 1:
How can index numbers be used to clarify the trend comparison between the following series of numbers?

Year	A	B
0	3,419	1,011
1	3,724	1,235
2	4,216	1,364
3	4,103	1,415
4	4,787	1,687

Answer:
Two more columns could be added with each number shown as a percentage of the first year, as follows:

Appendix **219**

			Index	
Year	A	B	A	B
0	3,419	1,011	100	100
1	3,724	1,235	109	122
2	4,216	1,364	123	135
3	4,103	1,415	120	140
4	4,787	1,687	140	167

Exercise 2:
In using index numbers, as in the foregoing example, what caution should be observed?

Answer:
If the base period is abnormal for either series, it will distort the relative numbers accordingly.

Exercise 3:
In the Consumer Price Index of the U.S. Bureau of Labor Statistics, what is used to weight the various prices?

Answer:
The quantity of the given item consumed in a typical household budget.

Exercise 4:
Does the Consumer Price Index measure changes in cost of living?

Answer:
No. From a practical viewpoint, there is no real way to define, much less measure, cost of living on any kind of truly comparable basis from year to year.

Exercise 5:
If different weights are applied, index numbers of the same original data will differ. How can this be?

Answer:
Each set of weights, in effect, is asking a different question of the data, and it is to be expected that the answers will differ accordingly.

Exercise 6:
If the two indexes in our first example were to be combined, with A being given twice the weight of B, what would be the combined index?

Answer:
The combined index would be as follows:

Year	A	B	Combined
0	100	100	100
1	109	122	113
2	123	135	127
3	121	140	127
4	140	167	149

Exercise 7:
How would you deflate the data shown below with the given price index?

Year	Price index	Data
0	123.7	2,050
1	127.4	2,101
2	131.3	2,142
3	135.7	2,142

Answer:
With the price index recalculated so that Year zero = 100, the price deflated data would be as follows:

Year	Price index	Data	Price deflated
0	100.0	2,050	2,050
1	103.0	2,101	2,040
2	106.1	2,142	2,019
3	109.7	2,203	2,008

STATISTICS AND SCIENTIFIC METHOD

Exercise 1:
What is the essence of scientific method in its approach to solving problems?

Answer:
If from the observation of data there is an apparent relationship of variables, the next step is to test this hypothesis by other data accumulated for that purpose to see if it holds there as well.

Appendix **221**

Exercise 2:

If a certain relationship between variables is first noted in a given set of data, can those same data be used to test whether or not the relationship does in fact exist?

Answer:

No. The relationship may be a quirk of the original data, and thus it must be tested with additional data in order to place any faith in the verification process.

Exercise 3:

If the hypothesis does hold true with additional data testing, does that mean that it can be relied on with total confidence?

Answer:

When the research is in certain fields such as physics, where outside variables can be controlled, great confidence can be placed in such results. But in the economic world, where so many variables are always uncontrolled, there is always much less than total confidence. Even so, the hypothesis may well serve as a practical working tool until such time as it is negated by other data.

Exercise 4:

Does scientific method have any practical value in the business world?

Answer:

There is always an effort to find cause and effect relationships in many situations, and the basic concept of scientific method is very useful in the discovery, testing, and utilization of such relationships.

INDEX

Acquisitions and internal rate of return, 67
Advertising recall and statistical inference, 83
Analysis of variance (ANOVA)
  concept of, 120–124
  exercises in, 210–211
Arithmetic average and frequency distributions, 10–11
Arithmetic line chart
  defined, 20–21
  examples of, 23–24, 30–33, 35
Arithmetic mean
  calculation of, 41
  defined, 36–37
  deviations from, 49–50, 52
  versus the geometric mean, 39–40
  and normal distributions, 86–88, 96, 100
  in survey interpretation, 101
  uses of, 52
Average deviation, 49–50
Averages, 36–42
  arithmetic, 10–11
  root mean square, 50–51
  unweighted, 44
  uses of, 36
  and variability, 47
  weighted, 43–46, 195–196

Band charts, 24
Bar charts
  base line of, 18, 20
  defined, 17–20, 35
  examples of, 23–25, 27–34
  histograms as, 13, 15–16
  three-dimensional, 23

BASIC computer language, 184–192
Binomial distribution, 85–96

Calculators (hand)
  for compounding, 58–60, 159
  for computation of internal rate of return, 69
  for computation of standard deviation, 51
  in regression analysis, 139
Capital outlays and internal rate of return, 67, 70
Central limit theorem, 100
Chain letters, 60
Change in mix, 44–46
Charts, 17–35 (see also Bar charts; Line charts)
  band, 24
  computer-generated, 21
  embellishment of, 21–22
  examples of, 23-25
  pie, 17–18, 22, 25, 35
  purpose of, 17, 22
  three-dimensional, 21–23, 25
  types of, 17–22
Chi square, 114–119
  calculation of, 115–116
  examples of, 116–119
  exercises in, 208–210
Class intervals, 10–13, 15–16
Cluster sampling, 110
Coefficient of correlation, 132
Coefficient of determination, 131–132
Coefficient of variation, 52
Coefficients in linear equations, 125–128

Combinations
  applications of, 81–82
  defined, 78, 80–81
  exercises in, 203–204
Compound growth rate
  exercises in, 217–218
  explained, 60–61, 156–161
Compounding
  calculated, 58–60
  exercises in, 199
  fraud in, 60–61
  in reverse, 60
Computers (personal)
  for computation of internal rate of return, 69
  for computation of standard deviation, 51
  for deriving seasonal indexes, 155
  in finding permutations and combinations, 81
  as generators of charts, 21
  language of, 184
  as organizers of data, 1
  programming of, 183–192
  in regression analysis, 139
  for telephone surveys, 109
  value of, 183, 192
Conference Board, the, 23–34
Consumer preferences and statistical inference, 83
Consumer Price Index, 165–168
Contingent statement, 186
Correlation (*see* Regression analysis)
Cost-of-living adjustments (COLAs), 165
Current ratio, 54
Curvilinear relationships, 133–134, 139

Data
  collection of, 2–4, 6–8
  comparison of, after seasonal adjustment, 150–152
  organization of, 1–8
  relevant versus irrelevant, 6–8
  sources of, 7–8, 70
Data organization, 1–8
  array concept in, 1, 9
  with computers, 1
  as first step in statistical analysis, 1, 8
  illustrated, 1–5
  rifle-shot approach to, 2–3, 6, 8
  shotgun approach to, 2–3, 6, 8
Debt ratio, 54
Decile values, 47–48
Degrees of freedom
  in analysis of variance, 123–124
  in chi square procedure, 116, 118
  in regression analysis, 131
Dependent variable in linear equations, 125, 133
Dichotomous population, 85

Empirical laws, 173
Experimental design, principles of, 120, 124
Exponential series, 60

Factorial, 78
Frequency distributions, 9–16
  and arithmetic average, 10–11
  charts for, 13, 15
  class intervals and, 10–13
  complex analysis of, 13
  exercises in, 193–194
  illustrated, 9–10

Galton, Sir Francis, 125
Geometric mean, 38–40
Growth rate (*see* Compound growth rate)

Index

Harmonic mean, 37–38
Histograms, 13, 15–16

Independence (in probability), 72
Independent variable in linear equations, 125, 133
Index numbers
  in the Consumer Price Index, 165–168
  exercises in, 218–220
  geometric mean and, 39
  pitfalls of, 163
  purposes of, 162–165
Installments and present value analysis, 63–66
Insurance companies, 59–60
Intercept in linear equations, 126
Interest
  compounding, 58–59
  and present value analysis, 65
Internal rate of return
  calculation of, 69–70
  defined, 67–69
  exercises in, 200–202
  and present value analysis, 66, 68
Investments and internal rate of return, 67

Least squares procedure
  in compound growth rate calculations, 158–161
  in regression analysis, 129–133
  in statistical research, 177–178
  in trend analysis, 136–139
Line chart
  defined, 17, 22
  examples of, 23–35
  three-dimensional, 25
  types of, 19–21
Linear equations
  in compound growth rate calculations, 158–160
  in regression analysis, 125–133
  in trend analysis, 136–139
Logarithms in compound growth rate calculations, 158–159
  (*see also* Semilog charts)

Mail surveys, 111
Market research surveys
  and statistical inference, 173–182
  by telephone, 109–110
Market share and statistical inference, 83
Media General Financial Weekly, 13, 173
Mean (*see also* Arithmetic mean; Averages)
  geometric, 38–40
  harmonic, 37–38
Mean square, 123
Measures of central tendency, 42
  (*see also* Averages; Mean)
Median, 41–42
Mode, 40–41
Multiple regression, 133–134
Mutually exclusive occurrences, 72

Nonlinear relationships in regression analysis, 133–134
Nonresponse rate, 110–111
Normal curve, 86–88
Normal distribution
  and chi squares, 116
  and statistical inference, 86–96, 100
  and survey interpretation, 101–102, 105–106
  values for, 89–95
Null hypothesis
  and chi squares, 115
  defined, 97–99
  in survey interpretation, 103, 105

Paired samples, 105–106
P/E ratio, 54
Percentages
  exercises in, 198–199
  probability expressed as, 73
  uses of, 53–57
Percentile values, 48–49
Permutations
  defined, 78–80
  exercises in, 203–204
Pie charts, 17–18, 22, 25
Ponzi schemes, 61
Present value analysis
  calculation of, 62–63
  exercises in, 199–200
  with installment payments, 63–66
  and internal rate of return, 66, 68
  and tax consequences, 65–66
Probability
  addition law of, 71–72
  defined, 71
  determined by counting, 75–76
  and error frequency, 74–75
  exercises in, 202–203
  expressed as percent, 73
  multiplication law of, 72
  and risk, 71
  samples, 108
  of success, 73–74
Probability laws
  addition, 71–72
  and market research, 174–175
  multiplication, 72
  and sample size, 112–113
  and statistical inference, 83–88, 96–100
  and unusual events, 76–77
Profit margin, 54
Profit projections, 61
Public opinion polls
  and statistical inference, 83, 96, 98–99
  by telephone, 109–110

Quadratic equations in regression analysis, 133
Quality control and statistical inference, 83
Quartile values, 48
Quick ratio, 54

Random sampling
  and analysis of variance, 123–124
  clusters in, 110
  defined, 108
  and statistical inference, 99
  and stratified segments, 109
  and survey interpretation, 104
Range, 47
Rate of return, 70 (*see also* Internal rate of return)
Ratios
  in compound growth rates, 157
  exercises in, 198–199
  geometric mean and, 39
  uses of, 53–54, 57
Regression analysis
  applications of, 131–132, 135
  for compound growth rate, 158–159
  defined, 125–127
  equations in, 133–134
  exercises in, 211–213
  method of, 127–130
Return on equity, 13–14, 53
Revenue projections, 61
Richmond, Virginia, 174
Root mean square average, 50–51
Rounding numbers, 171–172

Sample size, 111–112
Sampling error and statistical inference, 97–99

Sampling process (*see also* Survey results, interpretation of)
  interpretation of results of, 101–107
  and statistical inference, 97–100
Sampling techniques
  described, 108–113
  exercises in, 206–208
Sampling variability and statistical inference, 97
Scientific method
  defined, 173
  exercises in, 220–221
  and real-world problems, 176
  and statistics, 173–182
Scientific notation and factorials, 79
Seasonal indexes, 145, 149–155
Seasonal variation, 140–155
  and data comparisons, 150, 152
  defined, 140–141
  exercises in, 215–217
  indexes for, 145, 149–155
  procedure for, 142–148
  special techniques for, 154
Semi-interquartile range, 48
Semilog charts
  defined, 20–21
  examples of, 26–32, 35
  and index numbers, 163
Significant figures, 169–172
Simultaneous equations in regression analysis, 134
Social security, 165
Standard deviation
  defined, 50–51
  and normal distribution, 86–88, 96–97, 100
  as percentage of arithmetic mean, 52
  in survey interpretation, 101, 104–106
  uses of, 52
Standard error of b, 132

Standard error of the difference, 102–105
Standard error of estimate, 131–132, 136
Standard error of the mean, 101–102, 104
Statistical inference, 83–100
  exercises in, 204–205
  and prediction, 173
  and probability, 83–88, 96–100
Statistical significance
  in analysis of variance, 123–124
  and chi squares, 118
  of a deviation, 96–97
  of survey results, 101–107
Stratified segments, 109
String statement, 184
Sum of square, 121–123
Survey results, interpretation of (*see also* Sampling process and Sampling techniques)
  discussed, 101–107
  exercises in, 205–206

Telephone surveys, 109–110
Trend analysis (*see also* Seasonal (variation)
  described, 136–139, 155
  exercises in, 213–215
Turnover ratio, 54

United States Bureau of the Census, 7
United States Bureau of Labor Statistics, 7, 165–167
Unweighted averages, 44

Variability
  described, 47–52
  exercises in, 196–198

Variance, analysis of
  concept of, 120–124
  exercises in, 210–211

Weighted average
  changes in, 44–46
  described, 43–46
  exercises in, 195–196
  illustrated, 43
  and unweighted average, 44

Word Jumble, the, 80

Zero coupon bonds, 59